Lecture Notes in Mathematics

A collection of informal reports and seminars
Edited by A. Dold, Heidelberg and B. Eckmann, Zürich

118

Proceedings of the 15th Scandinavian Congress Oslo 1968

Edited by K.E. Aubert and W. Ljunggren,
University of Oslo, Blindern, Oslo

Springer-Verlag
Berlin · Heidelberg · New York 1970

Present address of Prof. Aubert

Tufts University
Department of Mathematics
Medford, MA 02155/USA

The present volume contains the majority of the one-hour lectures given at the 15th Congress of Scandinavian Mathematicians held in Oslo, August 12-16, 1968 - together with the lectures given by Carleson and Wermer at the Seminar on Function Algebras held in conjunction with the Congress.

The one-hour lectures omitted from this volume are those of Fenchel, Helgason, Hörmander and Jacobinski, some of which have already been published elsewhere.

We should also mention that Professor Oystein Ore was going to give a one-hour lecture on the four-color problem, but unfortunately he died in Oslo during the Congress before the scheduled time of his lecture.

On behalf of the Congress organizers, we wish to thank Springer-Verlag for their willingness to publish these talks as well as for their help with the preparation of manuscripts.

W. Ljunggren K.E. Aubert

Contents

Cassels, J.W.S.: Factorization of Polynomials in Several Variables 1

Dahl, O.J. : Programming Languages as Tools for the Formulation of Concepts 18

Fenstad, J.E. : Non-Standard Models for Arithmetic and Analysis 30

Hilton, P. : On Factorization of Manifolds 48

Lehto, O. : Homeomorphisms with a Given Dilatation ... 58

Martens, H.H. : From the Classical Theory of Jacobian Varieties 74

Selberg, A. : Recent Developments in the Theory of Discontinuous Groups of Motions of Symmetric Spaces 99

Carleson, L. : The Corona Theorem 121

Wermer, J. : Polynomial Approximation 133

FACTORIZATION OF POLYNOMIALS IN SEVERAL VARIABLES

J.W.S. Cassels

1. Today I want to talk about a concrete old-fashioned elementary algebraic problem. Let $f(x)$ and $g(y)$ be polynomials in the single independent variables x and y with coefficients in, say, the field $\mathbb{C}$ of complex numbers:

$$f(x) \in \mathbb{C}[x] \;,\; g(y) \in \mathbb{C}[y] \;. \tag{1.1}$$

When is

$$f(x) - g(y) \tag{1.2}$$

reducible considered as a polynomial in the pair of variables x,y ?

A trivial case is when f and g are the same polynomial since notoriously $f(x) - f(y)$ is divisible by $x-y$. More generally, if

$$f(x) = H(F(x)) \;,\; g(y) = H(G(y)) \tag{1.3}$$

for some polynomials $F,G,H,$ then $f(x) - g(y)$ is divisible by $F(x) - G(y)$. Another case of reducibility was discovered by Davenport, Lewis and Schinzel [1]. Let

$$T_4(z) = \cos(\arccos(z)) = 8z^4 - 8z^2 + 1 \tag{1.4}$$

be the 4th Čebyšev polynomial. If

$$f(x) = T_4(x) \;,\; g(y) = -T_4(y) \tag{1.5}$$

we have *

$$\begin{aligned} f(x) - g(y) &= 2(2x^2+2y^2-1)^2 - 16x^2y^2 \\ &= \{2^{\frac{1}{2}}(2x^2+2y^2-1) + 4xy\}\{2^{\frac{1}{2}}(2x^2+2y^2-1) - 4xy\} \end{aligned} \tag{1.6}$$

---

* One can get reducibility over the rationals by sacrificing symmetry in x and y and replacing y by $2^{\frac{1}{2}}y$.

One can show by a variety of ways (which are left to the ingenuity of the hearer that (1.5) is not a special case of (1.3). More generally $f(x) - g(y)$ is clearly reducible if

$$f(x) = T_4(F(x)) \ , \ g(y) = -T_4(G(y)) \tag{1.7}$$

for any polynomials $F(x)$ and $G(y)$. Davenport, Lewis and Schinzel were rash enough to conjecture that (1.3) and (1.5) are the only cases of reducibility but were able only to establish some fairly elaborate and apparently rather weak criteria which ensure irreducibility.

Somewhat earlier Ehrenfeucht [3] had established the following straightforward criterion:

THEOREM (Ehrenfeucht). Suppose that the degrees of $f(x)$ and $g(y)$ are coprime. Then $f(x) - g(y)$ is irreducible.

I shall not give Ehrenfeucht's simple proof here, as we shall obtain a different one later. I should, however, like to mention a couple of deductions that have been made from it.

In the first place Ehrenfeucht and Pełczyński (see Schinzel [4]) disposed completely of the analogous problem for more than two variables:

THEOREM (Ehrenfeucht and Pełczyński). Let $n > 2$ and let $f_j(x_j)$ $(1 \leq j \leq n)$ be nonconstant polynomials in the independent variables x_j. Then

$$\Sigma f_j(x_j) \tag{1.8}$$

is irreducible.

I sketch the elegant proof for the case $n = 3$, so suppose, with the obvious change of notation, that

$$f(x) + g(y) + h(z) = p(x,y,z)\, q(x,y,z) \ ,$$

where neither of the polynomials p,q is a constant. It is easy to see that we can find polynomials

$$Y(t)\ ,\ Z(t)$$

in a new variable t such that

(i) the degree of

$$e(t)\ (\text{say}) = g(Y(t) + h(Z(t) \tag{1.9}$$

is prime to the degree of f :

and (ii) The polynomials

$$p(x,Y(t),Z(t)\ ,\ q(x,Y(t),Z(t) \tag{1.10}$$

in x,t are not constants * .

---

* For let l, m, n be the degrees of f, g, h and let $k > 0$ be an integer. Take $Y(t)$, $Z(t)$ of degree kln , klm respectively. If we choose Y,Z as is obviously possible, so that $e(t)$ in (1.9) has degree $klmn - 1$, then (i) is satisfied. There is obviously so much freedom left that we can satisfy (ii) as well.

When I lectured in London on this work, Dr J.A. Tyrrel pointed out that the theorem of Ehrenfeucht and Pełczyński can also be proved by the methods of algebraic geometry. The points of intersection of $p(x,y,z) = 0$ and $q(x,y,z) = 0$ are singular points of the surface (*) $f(x) + g(y) + h(z) = 0$, so the surface has a singular curve. On the other hand the usual differential criterion shows that the surface (*) can have at worst isolated singular points, except, possibly "at infinity". And the singular points at infinity are also isolated if we suppose that f,g,h have the same degree. This is no loss of generality since we may make the substitution $x = x_1^{mn}$, $y = y_1^{nl}$, $z = z_1^{lm}$ with new variables x_1,y_1,z_1 .

But this is a contradiction, since $f(x) + e(t)$ is irreducible by Ehrenfeucht's criterion.

Some more general results of a similar type have been proved by Davenport and Schinzel [2] and by Schinzel [5]. There is an account of related problems (some of which have since been solved) in Schinzel [4].

The other application of Ehrenfeucht's Theorem that I want to mention is a rather curious mixed result of Schinzel's [6].

THEOREM (Schinzel). Suppose that $f(x)$, $g(y)$ have rational coefficients and that the degree of $f(x)$ is a prime, p, say. Then over the complex numbers either $f(x) - g(y)$ is irreducible or

$$g(y) = f(G(y)) \tag{1.11}$$

for some $G(y) \in \mathbf{C}[y]$.

For suppose that $f(x) - g(y)$ is reducible. Without loss of generality the highest coefficient in $f(x)$ is 1 and then we can normalize the factorization

$$f(x) - g(y) = H_1(x,y) \dots H_r(x,y) \tag{1.12}$$

into irreducibles of $\mathbf{C}[x,y]$ so that the coefficients of the highest occurring powers of x in each of the $H_j(x,y)$ is also 1. With this normalization, the $H_j(x,y)$ are uniquely determined up to order, and so are actually in $K[x,y]$, where K is some galois extension of the rationals. The automorphisms of K just permute the $H_j(x,y)$ amongst themselves.

By Ehrenfeucht's criterion, the degree n of $g(y)$ must be divisible by the prime degree p of $f(x)$, say

$$n = pu \tag{1.13}$$

Schinzel now considers the terms of highest weight in each of the terms of (1.12) when x is given weight 1 and y has weight u.

This gives

$$x^p - by^{pu} = L_1(x,y) \dots L_r(x,y) , \qquad (1.14)$$

where b is the coefficient of y^{pu} in $g(y)$ and where $L_j(x,y)$ consists of the terms of greatest weight in $H_j(x,y)$ and so is a form in x and y^u. The L_j are permuted by the action of the galois group and a discussion of the possible actions, which I shall not reproduce in detail, shows that at least one of the L_j is linear in x, say L_1. Then H_1 is linear in x, say

$$H_1(x,y) = x - G(y) . \qquad (1.15)$$

But now (1.12) immediately implies (1.11).

Schinzel also discusses conditions under which the coefficients of G are necessarily rational.

2. A year or two ago my pupil M.J.T. Guy and I were led to consider this problem of the reducibility of

$$f(x) - g(y) . \qquad (2.1)$$

Apart from a dramatic intervention by Bryan Birch, which will be narrated in its place, all the work I shall be describing is joint work with Guy. *

The first thing that we observed is that it is essentially not a problem of commutative algebra at all but one of combinatorial group theory. I want first to explain how this comes about.

We can write (2.1) in the shape

$$f(x) = t , \quad g(y) = t \qquad (2.2)$$

---

* During the conference I learned that H. Tverberg has proved some interesting results which overlap with ours. His methods are completely different.

for some new variable t . For the moment let us consider only one of these equations, say

$$f(x) = t \; . \tag{2.3}$$

From the point of view of complex variable theory this represents the Gauss sphere of x as a Riemann surface over the Gauss sphere of t . Let

$$t_0 = \infty \; , \; t_1 \; , \ldots , \; t_r \tag{2.4}$$

be the points of ramification. As usual we cut the Riemann surface into sheets by joining the ramification points by an arc on the t-sphere:

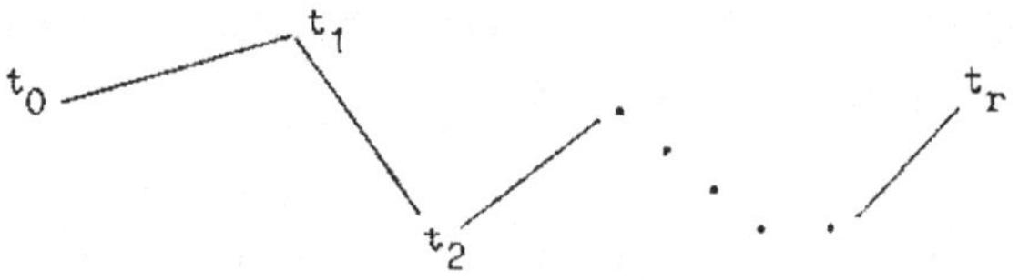

A circuit round t_s in a positive direction induces a permutation π_s (say) of the sheets. We can mumber the sheets so that the permutation π_0 corresponding to $t_0 = \infty$ is

$$\pi_0 \colon \; 1 \longrightarrow 2 \; , \; 2 \longrightarrow 3 \; , \ldots , \; m - 1 \longrightarrow m \; , \; m \longrightarrow 1 \, , \tag{2.5}$$

where m is the degree of f . As usual we write (2.5) in the form:

$$\pi_0 = (12 \; \ldots \; m) \; . \tag{2.6}$$

Since a circuit of all the ramification points leaves everything unaltered, we have

$$\pi_0 \, \pi_1 \; \ldots \; \pi_r = 1 \quad \text{(the identity permutation)} \tag{2.7}$$

A further heavy restriction on the π_s is given by the fact that the Riemann surface, being the Gauss sphere of x , has topological genus zero. In detail this condition works out as follows. As everyone knows, a permutation π of the number $1,2,\ldots,m$ can be expressed uniquely as the product of disjoint cycles, say of lengths

$$m_1,\dots,m_v \ , \ m_1+\dots+m_v = m \ . \tag{2.8}$$

We define the length of π to be

$$l(\pi) = \sum_{1\leq j\leq v} (m_j-1) \ . \tag{2.9}$$

Then the formula for the genus of a Riemann surface gives

$$\sum_s l(\pi_s) = 2(m-1) \tag{2.10}$$

or, equivalently,

$$\sum_{s\neq 0} l(\pi_s) = m-1 \tag{2.11}$$

since

$$l(\pi_0) = l((12\ \dots\ m)) = m-1 \ . \tag{2.12}$$

There is another way of looking at the length of a permutation which shows how restrictive the condition (2.11) is. One sees without trouble that $l(\pi)$ is just the minimum number of transpositions required to give π ; and this implies that

$$l(\pi\pi') \leq l(\pi)l(\pi') \tag{2.13}$$

for any two permutations π and π' . The equation (2.7) and repeated application of (2.13) thus give

$$\begin{aligned} l(\pi_0) &= l(\pi_0^{-1}) = l(\pi_1\ \dots\ \pi_r) \\ &\leq l(\pi_1)\ \dots\ l(\pi_r) \end{aligned} \tag{2.14}$$

The condition (2.12) now states that equality must hold here, and so in each of the applications of (2.13). In particular we must have

$$l\Big(\prod_{s_1\leq s\leq s_2} \pi_s\Big) = \sum_{s_1\leq s\leq s_2} l(\pi_s) \tag{2.15}$$

for any s_1, s_2 with

$$1 \leq s_1 < s_2 \leq r \tag{2.16}$$

We can go in the opposite direction. Given any distinct points $t_0 = \infty, t_1, \ldots, t_d$ and permutations $\pi_0, \pi_1, \ldots, \pi_d$ satisfying (2.6), (2.7) and (2.11), there is a polynomial $f(x)$ of degree m from which they can be obtained in the way just described. For these data define an abstract Riemann surface of genus zero. The Riemann Mapping Theorem then gives us functions on the surface, and, in particular, a function x having as its only singularity a simple pole at the unique point of the surface over $t = \infty$. It then follows readily from the monodromy theorem * that $f(x) = t$ for some polynomial f.

We note that the conditions above are entirely independent of the positions of the ramification points, provided that they are distinct. We can even make a pair of the finite ramification points coincide in a meaningful way. Suppose, for example, that we make t_{r-1}, t_r coincide at t'_{r-1}. Consider the following system of ramification points and associated permutations:

$$\begin{array}{llll} t_0 = \infty & t_1 & t_{r-2} & t'_{r-1} \\ \pi_0 & \pi_1 & \pi_{r-2} & \pi'_{r-1} = \pi_{r-1}\pi_r \end{array} \tag{2.17}$$

By (2.15) we have

$$l(\pi'_{r-1}) = l(\pi_{r-1}) + l(\pi_r) .$$

It is now readily verified that the analogues of (2.6), (2.7) and (2.11) hold for the new system. It thus corresponds to a polynomial. We shall speak of this as a confluent case.

---

* The elementary symmetric functions of the values of x over a given t are meromorphic functions of t. All except the product are everywhere regular; and the product has a simple pole at infinity as its only singularity.

3. Now let us return to the pair of equations

$$f(x) = t, g(y) = t \tag{3.1}$$

and, changing the notation slightly denote by

$$t_0 = \infty, t_1, \ldots, t_r \tag{3.2}$$

the values of t which are ramification points for at least one of f, g. As above, we obtain permutations

$$\pi_0 = (1,2,\ldots,m), \pi_1, \ldots, \pi_r \tag{3.3}$$

the sole difference being that now π_s is allowed to be the identity permutation (as happens when t_s is not a ramification point of f). Similarly g gives us permutations

$$\sigma_0 = (1,\ldots,n), \sigma_1, \ldots, \sigma_r \tag{3.4}$$

of $1,2,\ldots,n$, where n is the degree of g.

We now construct an abstract mn-sheeted Riemann surface Σ over the t-sphere on which both x and y are one-valued (the fibre product of the surfaces of f and g). The ramification points are the t_s and the sheets will be labeled by pairs

$$(u,v) \qquad (1 \leq u \leq m,\ 1 \leq v \leq n)\ , \tag{3.5}$$

The permutation of the sheets (3.5) at t_s will be $\prod_s$ defined by

$$\prod_s(u,v) = (\pi_s u, \sigma_s v)\ . \tag{3.6}$$

I assert that a necessary and sufficient condition for

$$f(x) - g(y) \tag{3.7}$$

to be irreducible is that the surface Σ just constructed be connected. For suppose, first, that

$$f(x) - g(y) = p(x,y)q(x,y) \tag{3.8}$$

is reducible. Then the pairs of branches of x,y over the same t which satisfy $p(x,y) = 0$ clearly give a proper open and closed subset Σ, of Σ. Conversely, if Σ_1 is a proper component of Σ, the monodromy principle gives us an

$$h(x,y) \in \mathbb{C}[x](y) \tag{3.9}$$

which vanishes on Σ_1 and which is of lower degree in x than is $f(x)$. This implies that (3.7) is reducible.

The condition that Σ be connected is just that the Π_s act transitively on the (u,v) and so can be expressed in terms of the permutations π_s and σ_s only. All the other aspects of the problem similarly have a simple interpretation in terms of the π_s and σ_s. As promised we have thus made our problem into one of combinatorial group theory.

The theorem of Ehrenfeucht which I discussed at the beginning of this lecture is an immediate consequence of what has just been said. For if the degrees m and n of f and g are coprime, then the permutation Π_0 at infinity already acts transitively on all the sheets (3.5).

We also note the rather surprising fact that the ramification points t_s have disappeared entirely from our analysis.

4. It is natural first to look at the case when there are just two finite ramification points, say

$$\begin{array}{llll} & t_0 = \infty & t_1 & t_2 \\ f: & \pi_0 & \pi_1 & \pi_2 \\ g: & \sigma_0 & \sigma_1 & \sigma_2 \end{array} \tag{4.1}$$

and to restrict attention to the case when *

$$\deg(f) = \deg(g) = n \quad \text{(say)}. \tag{4.2}$$

A simple case of reducibility is when the permutations are given by

$$\begin{aligned} \pi_0 &: u \to u+1 \quad ; \sigma_0 : v \to v+1 \\ \pi_1 &: u \to n-u \quad ; \sigma_1 : v \to n+1-v \\ \pi_2 &: u \to n-1-u \; ; \sigma_2 : v \to n-v \end{aligned} \tag{4.3}$$

where (now and henceforth) all indices are taken modulo n. It is immediately obvious that

$$(u-v) \to 1 - (u-v) \tag{4.4}$$

under the action of both Π_1 and Π_2. Hence the Riemann surface Σ for $n > 2$ splits into pieces Σ_w (say) where Σ_w consists of the sheets for which

$$\text{either} \quad u - v \equiv w \qquad \text{or} \qquad u - v \equiv 1 - w . \tag{4.5}$$

---

* if $f(x) - g(y)$ is reducible, it seems plausible that $f(x) = U(F(x)), g(y) = V(G(y))$ for some polynomials U,V,F,G such that

$$\deg(U) = \deg(V) = \text{g.c.d.}(\deg(f),\deg(g)$$

and

$$U(X) - V(Y)$$

is reducible. But we have not succeeded in finding a proof. (added at the conference). Tverberg showed that if there is a reducible $f(x) - g(y)$ which is not trivial or obtained by substitution from the T_4 case or the degree 7 case (see below) then there is also one in which the degrees of f and g are equal, namely $f_1(x) - g_1(y)$, where $f_1(x) = f(x^{2n} + x^{2n-1})$; $g_1(y) = g(y^{2m} + y^{2m-1})$.

The case $n = 4$ is just the new case

$$f(x) = T_4(x) \ , \ g(y) = -T_4(y) \tag{4.6}$$

with the Čebyšev polynomial T_4 , which was discovered by Davenport, Lewis and Schinzel, and which we have already discussed. Disappointingly, the remaining n give nothing essentially new, only f and g of one of the two trivial types *

$$f(x) = H(F(x)) \ , \ g(y) = H(G(y))$$

or (4.7)

$$f(x) = T_4(F(x)) \ , \ g(y) = -T_4(G(y))$$

---

* With appropriate positions of the ramification points (4.3) gives the Čebyšev polynomials $f(x) = T_n(x)$, $g(y) = -T_n(y)$ and Davenport, Lewis and Schinzel [1] remarked already that these give (4.7). It is more in the spirit of this lecture to argue directly from the permutations using the two following facts:

(i) If n is odd, and v_0 is defined by $2v_0 \equiv 1$, the substitution $v' = v - v_0$ makes it clear that f,g are essentially the same polynomial.

(ii) Suppose that $N|n$. Then in an obvious notation the surfaces belonging to f_n, g_n are coverings of the surfaces of f_N, g_N respectively, the coverings being given by reading (4.3) modulo N instead of modulo n . In terms of polynomials this just means that

$$f_n(x) = f_N(F(x)) \ , \ g_n(x) = g_N(G(x))$$

for some polynomials F,G .

Hence if n is divisible by an odd $N \neq 1$, we have the first half of (4.7) and if n is divisible by 4 we have the second half of (4.7). And if n is divisible by 4 but is not a power of 2 we can put f and g into both of the forms of (4.7).

Still confining ourselves to two finite ramification points and equal degrees, we made an exhaustive computer search up to $n = 12$. We did not go any further, partly because the number of relevant permutations starts to increase very rapidly but mainly because the computer we were using was murdered so that its memory could be used for another computer and we were too lazy to rewrite the program in a language that the cannibal computer could comprehend. The search threw up several new solutions for $n = 7$ and a new solution for $n = 11$. Further examination showed that the solutions for $n = 7$ could all be derived by confluence from a single solution with three finite points of ramification. We thus landed up with the two new solutions:

<u>$n = 7$</u>

	t_0	t_1	t_2	t_3
f	(1234567)	(65)(41)	(64)(32)	(76)(31)
g	(1234567)	(76)(41)	(54)(31)	(75)(21)

Here one component of Σ is given by

$v - u \equiv 0,2$ or $6 \mod 7$.

<u>$n = 11$</u> (write T for ten and E for eleven)

	t_0	t_1	t_2
f	(123456789TE)	(T3)(96)(54)(21)	(ET2)(953)(876)
g	(123456789TE)	(E4)(T9)(85)(21)	(E32)(T84)(765)

Here one component of Σ is given by

$v - u \equiv 0,4,5,6$ or 8.

All that remains to do is to find the corresponding polynomials. This is easier said than done, since the Riemann Mapping Theorem, which permits us to deduce the existence of f and g, is notoriously a pure existence theorem. What is clear is that the polynomials of degree 7 will have to contain essentially a free parameter, since there are three ramification points and a transformation of the type $t \to At + B$. $(A,B \in \mathbb{C})$ allows us to take only two of them into fixed points.

We were wondering how best to attack this problem when Bryan Birch happened to visit Cambridge. One of us mentioned the difficulty to him over lunch and before the lunch was over he had produced the requisite polynomials for $n = 7$, apparently by entirely low-brow fiddling. They are *

$$\begin{aligned} f(x) - g(y) &= x^7-7\lambda tx^5+(4-\lambda)tx^4+(14\lambda-35)t^2x^3- \\ &\quad -(8\lambda+10)t^2x^2+((3-\lambda)t^2+7(3\lambda+2)t^3)x- \\ &\quad -y^7+7\mu ty^5+(4-\mu)ty^4-(14\mu-35)t^2y^3- \\ &\quad -(8\mu+10)t^2y^2-((3-\mu)t^2+7(3\mu+2)t^3)y-7t^3 \\ &= [x^3+\lambda x^3y-\mu xy^2-y^3-(3\lambda+2)tx+(3\mu+2)ty+t]\times \\ &\quad \times[x^4-\lambda x^3y-x^2y^2-\mu xy^3+y^4+2(\mu-\lambda)tx^2-7txy+ \\ &\quad +2(\lambda-\mu)ty^2+(3-\lambda)tx-(3-\mu)ty-7t^2]. \end{aligned}$$

where t is a parameter and

$$\lambda = \tfrac{1}{2}(1+\sqrt{(-7)}) \ , \ \mu = \tfrac{1}{2}(1-\sqrt{(-7)}) \ .$$

---

* The forms of degree 7 were obtained by Tverberg independently. He showed that if $f(x) - g(y)$ contains a factor of degree 3, then it is either trivial or got by substitution from the degree 7 case.

He also showed that if $f(x) - g(y)$ has a quadratic factor it is either trivial or obtained by substitution from the Davenport-Lewis-Schinzel T_4 case.

Not content with this, over dinner he similarly produced the polynomials for $n = 11$. They are

$$
\begin{aligned}
& f(x) - g(y) = \\
& x^{11}+11(\lambda,-2,-3\mu\tau,-16\lambda,3\mu^2(\lambda-4),30\mu\tau,-63\mu,\\
&\qquad\qquad -20\mu^4,3\mu^4\tau^2,-9\theta)(x,1)^9-\\
& -y^{11}-11(\mu,-2,-3\lambda\sigma,-16\mu,3\lambda^2(\mu-4),30\lambda\sigma,-63\lambda,\\
&\qquad\qquad -20\lambda^4,3\lambda^4\sigma^2,9\theta)(y,1)^9\\
& = [(1,-\lambda,-1,1,\mu,-1)(x,y)^5+\theta(2,-\lambda,-\mu,2)(x,y)^3-\\
&\qquad -2\theta(\mu,-3,\lambda)(x,y)^2+\theta(\mu^3,\lambda^3)(x,y)-6\theta]\times\\
& \times[(1,\lambda,\sigma,2,\tau,\mu,1)(x,y)^6+\theta(\mu\tau,-\lambda^3,-2\theta,\mu^3,-\lambda\sigma)(x,y)^4+\\
& +2\theta(\lambda,\lambda^2,-\mu^2,-\mu)(x,y)^3-\theta(\mu(2\theta+3),3\theta,\lambda(2\theta-3))(x,y)^2+\\
&\qquad +4\theta(-\mu^3,\lambda^3)(x,y)+33],
\end{aligned}
$$

where

$$\theta^2 = -11, \quad \lambda = \frac{-1+\theta}{2}, \quad \mu = \frac{-1-\theta}{2},$$

$$\sigma = \mu-1, \quad \tau = \lambda-1 .$$

Of course the coefficients are irrational, in consonance with the theorem of Schinzel which I mentioned near the beginning of the lecture.

And this is where the matter rests. Guy conjectures that in fact there are now no further surprises, but I am more sceptical. We have not even been able to dispose of a possibility which Guy calls "Cassels' Monster".

As I explained earlier we can let the ramification points for a solution of the problem coalesce in various ways and still have a solution of the problem. So one could expect to pick up solutions with more than two finite ramification points from the two ramification points case, as indeed happened with us for $n = 7$. But there is the theoretical possibility, which we can see no way of excluding, that all the confluent cases of a nontrivial solution should turn out to be trivial. A study of the two-ramification case would then give no clue to its existence. Needless to say for the range

$$\deg(f) = \deg(g) \leq 12$$

one could find the "Monster", if it exists, by a suitable search, but we have not attempted to do this.

In conclusion, I note that our investigation has shed some negative light (if the expression is permissable) on another problem (for the background see e.g. Schinzel [4]). For what polynomials $f(x)$ is

$$e(x,y) = \frac{f(x) - f(y)}{x - y}$$

reducible? The known cases of reducibility are:

$$f(x) = H(F(x)) \qquad \deg(H) > 1 \ , \ \deg(F) > 1 \ ,$$

$$f(x) = (Ax+B)^n + C \ , \qquad A,B,C \in \mathbf{C}$$

and

$$f(x) = AT_n(Bx+C) + D \ , \qquad A,B,C,D \in \mathbf{C} \ ,$$

where T_n is the n-th Čebyšev polynomial. Our calculations show that there are no further cases with $\deg(f) \leq 12$ and such that $f(x) = t$ has only 2 finite ramifications.

There is clearly a great deal still to be found out.

R E F E R E N C E S

[1] H. Davenport, D.J. Lewis and A. Schinzel. Equations of the form $f(x) = g(y)$. Quarterly Journal (Oxford) (2), 12(1961), 304-312.

[2] H. Davenport and A. Schinzel. Two problems concerning polynomials. J. reine angew. Math. 214/5(1964), 386-391.

[3] A. Ehrenfeucht. Kryterium absolutnej nierozkładalności wielomianów. Prace Mat. 2(1958), 167-169.

[4] A. Schinzel. Some unsolved problems on polynomials. Neki nerešeni problemi u matematici. (Matematička biblioteka 25, Beograd, 1963).

[5] A. Schinzel. Reducibility of polynomials in several variables. Bull. de l'Acad. Polonaise des Sciences (Série des sci. math. astr. et phys.) 11(1963), 633-638.

[6] A. Schinzel. Reducibility of polynomials of the form $f(x) - g(y)$. Colloquium Math. 18(1967), 213-218.

Programming Languages as Tools for the Formulation of Concepts

Ole-Johan Dahl

The art of programming digital computers has something in common with the science of constructive mathematics. Then why is it called an art and not a science? Let us take a brief look at some of the similarities and differences between the two subjects.

Mathematics aims at clarity of thought and rigidity of proof. Any mathematical theory rests on a foundation of primitive concepts, axioms, and rules of inference. An important part of mathematical research is concerned with establishing such foundations as rigidly as possible, in order to provide secure bases for further thought. A second important aspect is the definition of fruitful composite concepts, which may act as backup for intuitive thinging and as application tools.

Programming, in a narrow sense, is the formulation of computing processes in terms of a given computer. When programming in machine language, the programmer must express himself in terms of the instruction code of the given computer, and take into account its memory structure and other relevant properties. The computer therefore, in a way, replaces the set of primitives and axioms accepted by a mathematician. The programmer is faced with an ultimate and inescapable check on his own clarity of thought: his program must be submitted to the computer for execution. Then it either works or does not work.

Thus, in art of machine programming one is not free to choose primitives and axioms at will, on the other hand, there is no need to search for clarity and rigidity. All these concepts are there - in the form of a computer.

A programmer is concerned with constructive concepts: discrete information structures and associated algorithms. Both must be finite and, furthermore, efficient with respect to storage space and computing time. From a practical point of view a computing process is uninteresting unless it is within the capacity of an available computer.

High level programming languages introduce notations and primitives which are to some extent machine independent. This is achieved partly by accepting computer primitives which are sufficiently similar on different computers (e.g. "real" numbers and arithmetic operations), partly by introducing language primitives which may correspond to compound concepts in computers. It is important that a programming language should present an essentially correct picture of computer capability, and thereby invite programmers to make efficient use of the computer. With this reservation in mind it should be the purpose of a programming language to make programming as easy as possible.

There are several aspects of this: to reduce the amount of irrelevant detail by making the computer itself take care of much of the internal administration, to protect the programmer against his own errors by consistency checks and logical elimination of certain kinds of errors, to provide special purpose concepts and programming aids which apply to special problem areas. For a general purpose programming language perhaps the most important aspect is to provide a framework for concept creation.

Electronic computers are designed to perform large scale information processing at fantastic speeds. The results which can be achieved are perhaps limited more by our own power to conceive and describe efficient computing processes, than by the power of computers to perform them. Now, how does the human mind react when faced with phenomena which are complicated in terms of the number of relevant details? Apparently by applying suitable concepts which somehow can organize the mass of details into understandable patterns. If suitable concepts for a given problem or problem area are either available in a programming language or may be formulated within it, then computation processes may be described directly in terms of them. In this way the programming language may become a tool of thought which is useful for problem analysis as well as computer programming.

Two fundamental principles of thought seem to be involved in the creation of concepts.

1. Decomposition. The human mind must concentrate. Otherwise coherent and precise thinking is impossible.

2. Classification. We have a need to generalise. A generalisation may be a statement about a class of things, and may thus amount to a large number of specific statements about the specific members of the class.

Any useful concept must have some degree of generality, which means that it is a class of specialized instances. It also must concern a limited aspect of reality obtained by decomposition.

In the following we shall look at some mechanisms of classification and strategies for decomposition found in programming languages. First we notice that both principles are related to the notion of programmed computers itself.

A computer program is physically a piece of text recorded on some medium. An execution of the program by a computer is a computing process. It is fruitful to regard a program as the class of its executions.

The nature of an automatic computer implies that a computing process is to a large extent decomposed from the environment of the computer. In fact, that is what the word "automatic" means. If the program determines every detail of the computing process, the decomposition is complete. In that case all executions of the program are in some sense equivalent. If, on the other hand, a computing process interacts with the environment, it is not completely determined by its program. Then the class of program executions may be a very wide one.

As one trivial example consider a program which is executed in the following steps:

1. Read a real number from an input medium;
2. Compute the square root of that number;
3. Write the result onto an output medium;
4. Stop.

For each execution of this program the details of the computing process will depend on the number read in. The class of executions, however, i.e. the program itself, represents the concept of taking the square root of a real number. The generality is obtained through interaction between the computing process and its environment, which is in this case an input medium external to the program.

Special purpose application programs, less trivial but essentially similar to the one above, exist in great numbers. Any computer installation will have a library of such programs available for use on the machine. A person may know nothing about computer programming and still make good use of a program library. From a language point of view, however, his situation is comparable to that of a two year old child, who has available a limited number of concepts and is able to utter the corresponding words, but still cannot make sentences.

In a machine language sentences are formed by stringing together machine instructions which perform elementary operations on elementary memory units. There is usually no inherent framework for constructing concepts which are usable directly within the language. However, some important semi-standardised notions of program decomposition have been developed, such as sub-routines, co-routines, and sequential processes in a multiprogramming environment. The formats of program components and the mechanisms for interaction and communication between them follow rules and conventions which are outside the language itself. As an example program components might communicate through a common memory area specially allocated for that purpose.

Incorrect interfacing of program components is one of the major sources of programming error. A high level programming language may provide a high degree of protection against such errors by including conventions for decomposition and interaction as part of its own structure.

As an example of a high level general purpose programming language we choose Algol 60. Decomposition and classification at the elementary level of Algol 60 correspond to features available in most computers. Its most important elementary concepts may be grouped under the following headings:

1. Value types. A value type is a class of elementary data items: real, integer, or Boolean. There are corresponding elementary operations: arithmetic operations and relations, logical composition and discrimination on truth values.
2. Variables. A variable represents the class of its potential values, which are elementary data items of a given type. Operations associated with a variable are: assigning and accessing its value.
3. Arrays. An array is a classification of a given number of variables. The variables are said to be the elements of the array. Associated with an array is the operation of subscripting for identifying an element of the array.

The central mechanism in Algol 60 of decomposition at a nonelementary level is the block concept. A block has the following format:

begin declarations; statements **end**

The declarations of a block identify and describe quantities such as variables, arrays, and possibly program components (procedures). The statements describe operations to be performed. The declarations introduce a nomenclature valid within the textual scope of the block, and referring to quantities which exist during an execution of the block. The quantities are said to be local to the block.

A block is itself a statement, and therefore it may be part of another block. The textual nesting of blocks may be carried out to any depth. As far as references to local quantities are concerned, a block is completely selfcontained and decomposed from its program environment; however, it may interact with the environment by referring to quantities local to enclosing blocks.

A statement represents the class of its potential executions. An execution of an Algol block, which is a statement, is called a "dynamic block instance", or simply a "block instance". A block, then, represents a class of potential block instances. Like a program as a whole a block is a piece of program text; a block instance, however, is a computin process which requires time for its operations and memory space for its local quantities. Several instances of a given block may co-exist as parts of an Algol program execution. Each of these block instances will have its own set of local quantities.

The Algol procedure concept is a formalisation of the informal notion "sub-routine". The body of a procedure is a block with certain additional interaction facilities. There is a calling mechanism which makes it possible to distinguish textually between the definition and the use of a procedure, and there are facilities for transmitting parameters and a function value at the time of its execution. These mechanisms serve to further decompose a block, i.e. a procedure body, form its program environment, or conversely, to widen the class of it potential executions through interaction with calling sequences.

Consider the following procedure, which represents the concept of computing the square root of a given positive number.

```
real procedure sqrt(x); value x; real x;

begin real y,z;

      y := (1+x)/2;

      for z := (y+x/y)/2 while z<y do y := z;

      sqrt := z

end
```

Although possibly embedded in a larger program the procedure is completely decomposed, in the sense that all quantities referred to are defined locally. The identifiers in use are "sqrt", "x", "y", and "z". Their defining occurrences are found on the two first lines of the declaration. Still an instance of the procedure is able to interact with its environment through bindings established by the calling mechanism. An example of a call is "sqrt(2)", which syntactically is an expression. Its evaluation is an instance of the procedure in which the parameter x has the value 2. This computing process delivers as its result a function value close to $\sqrt{2}$, which is by definition the value of the above expression. The computing process is a specialised instance of the more general concept "sqrt" defined by the procedure declaration.

The above procedure may be compared to the square root program described earlier. The big advantage of the formalised procedure mechanism is that specialised instances of the declared concept are meaningful syntactic units of the programming language, and may therefore be used as building blocks to compose programs.

Many Algol implementations permit full textual decomposition of procedure declarations, and have facilities for maintaining libraries of such procedures and facilities for programs to make reference to them. In this way the basic language may be augmented by special purpose concepts immediately available to a whole community of users.

Data and operations on data are strongly interrelated in our minds, so strongly, perhaps, that to consider the one is meaningless without also considering the other. The two are best thought of as complementary aspects of computing processes, the extensive and the intensive aspect by stretching an analogy.

The declarations and statements of a block correspond to the two aspects of a block instance, respectively a data structure with associated procedures, and a sequence of operations. This symmetry of the Algol block may account for some of its power as a basis for concept construction. However, in Algol 60 there is a strong emphasis on the intensive aspect of block instances. There is no mechanism for observing a block instance as something which is; one can only observe the effects of its operation. This emphasis on what is done, rather than what exists, has certain consequences for the kind of concepts that may easily be represented within Algol 60.

My further examples of decomposition and classification will be drawn from an extension of Algol 60 called Simula 67. The language has been developed at the Norwegian Computing Centre, Oslo.

One addition is a reference mechanism for naming block instances of a certain kind, called objects. There are also notations for getting access to the local quantities (variables, procedures, etc.) of named objects. These facilities make it possible to identify and observe objects; at the same time an object may be a "self-conscious" computing process in the Algol 60 sense, able to interact with its environment through non-local quantities, parameters, and in other ways.

Objects are members of classes, defined by class declarations. The larger part of a class declaration is a block called the class body. Objects of a given class are dynamic instances of the class body of that class declaration. It is a principle of Algol 60 that any number of instances of a given block may co-exist; therefore an object class may have any number of members in a Simula program execution.

As an example the following (partly informal) class declaration represents in a closed form the concept of "n-point Gauss integration", which is a technique of approximating definite integrals by weighted sums of function values.

```
class Gauss(n); integer n;

begin array x,w [1:n] ;

      real procedure integral (f,a,b);

            real procedure f; real a,b;

      begin integer i; real sum; sum := 0;

            for i := 1 step 1 until n do

            sum := sum + w[i]*f(a+(b-a)*x[i]);

            integral := (b-a)*sum

      end;

      compute the optimum weights, w[i], and abcissas,

      x[i], for the interval (0,1) as functions of n

end
```

An object of the class "Gauss" has the following local quantities: the integer "n" whose value is specified by parameter transmission, the arrays "x" and "w" whose values are computed once by the object itself, and the function procedure "integral" intended for repeated use from outside the object. Notice that the class declaration is fully decomposed in the sense that only local quantities are referred to in its body.

The following additional piece of program

```
ref (Gauss) G1,G2;

G1 :- new Gauss (5); G2 :- new Gauss (7);
```

will establish two objects, named G1 and G2, which represent specialised instances of the declared concept, respectively 5- and 7-point Gauss integration. (The sign ":-" represents the naming operation and is read "denotes".) The local quantities of the objects are available through a "dot" notation. For example

```
G1.integral(F,A,B)
```

is a call for the procedure declared within the object named "G1". The resulting value is therefore an approximation to the definite integral of the given function in the given interval, obtained by evaluating a 5-point Gauss formula.

The last principle of decomposition which I should like to sketch out is a consequence of a mechanism which I shall call "concatenation". Consider two blocks (or programs)

begin **D** a; $S_1(a,b)$ **end** and **begin** **D** b; $S_2(a,b)$ **end**

where "**D** a" stands for the declarations of a group of quantities collectively named "a", and "S(a,b)" stands for a sequence of statements containing occurrences of identifiers of the groups "a" and "b". We shall say that occurrences of identifiers a in S_1 are "bound" to defining occurrences in the local declarations, and that occurrences of b in S_1 are "free", and similarly for the second block. Algol 60 has the following standard "binding rule" for free identifier occurrences of a block: they are bound to defining occurrences, if any, of the smallest block enclosing the given one, otherwise they are taken as free occurrences of the latter.

The result of "concatenating" the two blocks above shall be a block of the form

```
begin D a; D b; S1(a,b); S2(a,b) end
```

which means that the following mechanisms are involved: juxtaposition of data structures in space, juxtaposition (in some way) of operations in time, and the binding of free identifier occurrences in one block to defining occurrences in the second and vice versa.

A concatenation mechanism permits the decomposition of programs into parts containing free identifiers. Suitably modified and restricted concatenation facilities are available sometimes in the "environments" of programming languages, such as machine (assembly) language and FORTRAN. A "loader" may be able to establish a complete program by bringing together components from different sources, and at the same binding "external" (i.e. free) identifiers.

In Simula 67 there is a formalised mechanism for the concatenation of class bodies. Concatenation is indicated by a prefix notation, and is asymmetric in the sense that a prefixed class declaration defines a "subclass" of the class identified by the prefix.

Example:

```
class A; begin D a; ..... end;

A class B; begin D b; ..... end;
```

The class B is a subclass of the class A. Its effective class body is the result of concatenating the (effective) body of A and the textual body of B. Consequently an object of the class B has all the properties (a) of a class A object and then some (b) in addition.

The prefix notation permits hierarchies of classes and subclasses to be defined. This has certain advantages with respect to the security and flexibility of naming operations. It also simplifies the problem of implementing efficiently the binding of free identifiers of the (textual) body of a subclass to defining occurrences in the (effective) body of the prefix class. The reverse bindings require certain additional specifications, and are defined so that declarations in a given class may be effectively replaced by declarations in a subclass.

The following is a simplified fragment of a class containing special purpose concepts for describing discrete event simulation models. The class is a standard part of the Simula 67 language (in the same sense as the procedures "sin" and "cos" are parts of Algol 60).

```
class SIMULATION;

begin class process;

      begin real evtime;
      .............
      end;

      ref(process) current;

      real procedure time; time := current.evtime;

      .............

      .............

end
```

Objects of a Simula program execution may interact in the mode of co-routines. This means that the operations of an object may be grouped together in "active phases" separated by inactive periods. One object is active at any one time, while others may be waiting for their next active phase to be invoked. A simulation model will contain objects belonging to (subclasses of) the class "process". At any time the active one will be named "current". Each process object will have a local variable "evtime" indicating the system time scheduled for the event of its next (or current) active phase. Consequently the procedure "time" gives access to the current value of the simulated time coordinate. There are procedures and other mechanisms for the sequencing of active phases in an order of non-decreasing event times.

There is in Simula 67 a prefix mechanism for concatenating class bodies and ordinary in-line blocks. One may regard it as a notation to establish a desired context of programming, where certain identifiers have predefined meanings as given by the declarations local to the prefix class body. The following fragment of a simulation model may serve as an example. The concepts "process", "current",and "time" are those defined in the class "SIMULATION".

```
SIMULATION begin process class person;

           begin real birth time;

                 real procedure age; age := time - birth time;

                 ........;

                 birth time := time; .........

           end;

           person class woman;

           begin ref(man) husband; Boolean yes;

                 ...........

           end;

           person class man;

           begin ref(woman) wife;

                 procedure marry(x); ref(woman) x;

                 if x.age<age ∧ x.yes then

                 begin wife :- x; x.husband :-current;

                       x.yes := false

                 end;

                 ............

           end;

           ............

           end;
```

Any "person" (man or woman) will set his (her) local variable "birth time" equal to the current system time during his (her) first active phase. Thereafter he or she has a well defined "age". A "man" will only decide to "marry" a "woman" younger than himself, and only if she says "yes".

Implementations of Simula 67 will have facilities for maintaining libraries of precompiled class declarations. Such "external" classes will represent concepts available to the user community. Highly structured classes, containing families of interrelated special purpose concepts, may be thought of as "application languages", which may be superimposed on top of the basic language through the mechanism of block prefixing. The standard class "SIMULATION" is one example of this.

Oslo, February 1969.

References:

1. P. Naur, ed: The Revised Report on the Algorithmic Language Algol 60.

2. O-J. Dahl, B. Myhrhaug, K. Nygaard: The Simula 67 Common Base Language. (Norwegian Computing Centre, Forskningsveien 1B, Oslo 3.)

NON-STANDARD MODELS FOR ARITHMETIC AND ANALYSIS*

Jens Erik Fenstad, University of Oslo

In this lecture I shall first present some of the basic ideas involved in the so-called "non-standard" theory of arithmetic and analysis. In particular I want to explain how this theory allows us to talk with preciseness about "infinitesimals", i.e. infinitely small, but non-zero numbers. The founders of analysis, in particular Leibniz, worked freely with "infinitesimals", and their use in the early stages of the development of analysis was, no doubt, of great heuristic importance in the discovery of new results. Later, as we all know, due to obvious inconsistencies, infinitesimals were bannished from decent mathematics, and replaced by the Weierstrassian ε - δ - method.

In the last part of the lecture I shall comment upon the relationship between the "non-standard" and the "standard" theory. Whereas the first part is purely expository, the last part may contain some new remarks.

1. ULTRAPRODUCTS AND NON-STANDARD MODELS FOR ARITHMETIC.

The non-standard theory presupposes a certain background from logic. The basic concepts in this connection are: <u>language</u>, <u>model</u>, and the notion of <u>satisfaction</u>.

"Language" is taken to mean a first order formalism with an arbitrary number of function and relation symbols. The important fact is that we have variables only for indiduals or elements so that the quantifiers $\forall$ and $\exists$ only can be applied to this kind of objects, and not to classes of elements, to functions and relations.

* The following is the text of a lecture given at the Opening Session of the Congress. The talk was therefore not aimed at specialists. It's purpose was to convey to the "general mathematician" some of the ideas involved in a not too technical fashion.

A language L is determined by its class of well-formed formulas. A formula Φ is by itself a syntactic object and has no meaning. But it can aquire a meaning by providing an interpretation of the language L. An interpretation is given by specifying a model M for L. A model (or structure) M consists of a non-empty collection of elements together with certain functions and relations defined in this set. (We assume that there is a one-to-one correspondence between function and relation symbols in L and functions and relations defined in M).

The basic concept is the notion of satisfaction, that the formula Φ from L is satisfied in M. A precise definition of this notion, and hence of the concepts of validity and truth, was given by A. TARSKI in the early nineteen thirties. It is interesting to note that his definition is an exact technical counterpart to the Aristotelian truth concept.

I do not have time to give the definition here. From an intuitive point of view the definition is quite obvious, and says nothing more than what every mathematician understand when he says that the commutative law (which is a formal statement $\Phi = \forall x \forall y\ (xoy = yox)$ in a suitable language L_G for group theory) is satisfied in a particular group G (which in one of the possible models for this language).

Within logic one has extensively studied various techniques of model constructions. One of the most powerful, which can be traced back to some work of T. SKOLEM in the middle thirties, is the so-called <u>ultraproduct</u> construction. An explicit form of this construction was first given by J. ŁOŚ in 1955, and it has since been refined and developed by the logic school under Tarski in Berkeley.

The ultraproduct is derived from the direct product. Let $\{M_i \mid i \in I\}$ be a family of models of a language L.. Form the direct product $M_0 = \prod_{i \in I} M_i$. Choose an ultrafilter D in the index set I, and introduce an equivalence relation in M_0 by requiring two functions $f, g \in M_0$ to be equivalent (in symbols: $f \tilde{D} g$) if they are equal for "almost all" $i \in I$, i.e. if $\{i \in I \mid f(i) = g(i)\} \in D$. (The "almost all" terminology makes good sense if one chooses an ultrafilter D which refines the filter of co-finite sets in I).

Let ${}^*M = \prod_{i\varepsilon I} M_i/_D$ be the set of equivalence classes with respect to $\underset{D}{\sim}$. We will in particular be interested in the case where each M_i is equal to some given model M. In this case we write ${}^*M = M^I/_D$.

Every relation and function in M can be extended to ${}^*M$: Let S be a binary relation in M, its extension ${}^*S$ is given by

$${}^*S(f/_D, g/_D) \Longleftrightarrow \{i \in I \mid S(f(i),g(i))\} \in D.$$

The following main result was first explicitly proved by J. Łoś, although it was implicit in the work of Skolem.

EXTENSION THEOREM. <u>A sentence Φ in L is valid in ${}^*M$ if and only if it is valid in M</u>.

Two models M_1 and M_2 of L are called <u>elementarily equivalent</u> if they admit exactly the same class of valid sentences of L. If at the same time M_1 is isomorphic to a submodel of M_2, M_2 is called an <u>elementary extension</u> of M_1.

Obviously the correspondence which maps every $x \in M$ on the constant function with value x, is an injection of M into ${}^*M$. The extension theorem says that <u>the ultrapower ${}^*M$ is always an elementary extension of M</u>.

The general extension principle is the most important aspect of the ultraproduct construction: ${}^*M$ is an extension of M, not necessarily isomorphic to M (which happens if D is a principal ultrafilter), but similar to M with respect to <u>all</u> properties expressible in the language L. If M is a group, then ${}^*M$ is also a group; if M is an integral domain, then so is ${}^*M$, etc. And the general extension principle can only be stated when one has the precise concepts of language, model, and satisfaction.

As an application of the ultraproduct construction we shall give the construction, dating back to Skolem's work, of non-standard models for arithmetic.

The language L_P in this case contains symbols for successor, addition, multiplication, and the less-than relation. It further contains names for the "ordinary" natural numbers. The standard model is N, i.e. the set of natural numbers with the usual arithmetic operations and relations.

Define ${}^*N = N^N/_D$, choosing the index set $I = N$. ${}^*N$ is elementary equivalent to N, i.e. an arithmetic sentence Φ in the language L_P is valid in N if and only if it is valid in ${}^*N$. If we let D be a free (i.e. non-principal) ultrafilter on N, then ${}^*N$ is not isomorphic to N: Since N is linearly ordered by the $<$ - relation, ${}^*N$ is also linearly ordered by the extension of $<$ to ${}^*N$. It is easy to see that N is an initial segment of ${}^*N$ in this ordering. Let f be the identity map from N to N. Then $\alpha = f/_D$ will be an "infinitely large" element in ${}^*N$, i.e. $n < \alpha$ for all $n \in N$.

One may show that the order type of ${}^*N$ is $\omega + (\omega^* + \omega) \cdot \lambda$, where λ is the type of a densly ordered set without a first or last element, and ω is the type of N, whereas ω^* is the converse type.

Our construction has the following somewhat unexpected consequence: There is no class of sentences in the language L_P which characterizes N up to isomorphism.

This is the famous result of Skolem on the non-characterizability of the natural numbers within a first order language.

2. NON-STANDARD MODELS FOR ANALYSIS: THE THEORY OF A. ROBINSON

Let R denote the real numbers, and consider the language L_R which contains names for all real numbers, and for all relations and functions definable over R. The language L_R will be rather rich in vocabulary, but this is of no concern to us here,- we are not in this lecture interested in any "constructive" justification of classical analysis.

Let ${}^*R$ be a non-standard extension of R. ${}^*R$ is an elementary extension of R. In L_R we have a predicate $N(x)$ which is satisfied in R by precisely the natural numbers N. $N(x)$ has also an interpretation in ${}^*R$, and it is easy to see that ${}^*N$, the interpretation in ${}^*R$, is a non-standard model of Peano arithmetic.

Seen from the outside ${}^*R$ is an ordered field which is a proper extension of R. ${}^*R$ is therefore a non-archemedian ordered field. (It is a hyperreal field in the terminology of Hewitt).

But note that the statement "For all reals a and b, such that $0 < a < b$, there exists a natural number n such that $b < n \cdot a$" is a sentence in L_R which is true in R, hence by the extension principle also true in ${}^*R$. This does not contradict the fact that ${}^*R$ is non-archemedian. For given elements $a,b \in {}^*R$ it may be necessary to choose an infinitely large number, i.e. some $n \in {}^*N - N$.

Since ${}^*R$ is non-archemedian we have non-empty sets

$$M_f = \{a \in {}^*R \mid |a| < r \quad \text{for some} \quad r \in R^+\}$$

$$M_i = \{a \in {}^*R \mid |a| < r \quad \text{for all} \quad r \in R^+\}.$$

M_f are the finite elements of ${}^*R$, M_i the infinitesional ones. The elements of R are called <u>standard</u>. The elements of ${}^*R-M_f$ are called <u>infinite</u>, and those of ${}^*R-R$ <u>non-standard</u>.

It is easy to see that M_f is a ring and that M_i is a maximal ideal in M_f; it follows that $M_f/_{M_i} \cong R$.

Within this frame-work we shall develop some of the basic ideas of non-standard analysis. We shall follow closely the expositions given by A. Robinson to whom the theory is due.

Our first topic is taken from the theory of sequences. A sequence $\langle s_n \mid n \in N\rangle$ can be considered as a function $s: N \to R$. It is thus represented by some function symbol s in L_R. In the transition form R to ${}^*R$ the standard function s will be extended to a function ${}^*s: {}^*N \to {}^*R$. This is an immediate consequence of the general extension principle: The statement Φ which describes s as a function from N to R is also valid in ${}^*R$ and describes there a function from ${}^*N$ to ${}^*R$. Write ${}^*s(\omega)$ as s_ω.

THEOREM. <u>For any standard sequence $\langle s_n \mid n \in N\rangle$, $\lim s_n = s$ in the usual sense, if and only of $s - s_\omega$ is infinitesimal for all $\omega \in {}^*N-N$.</u>

We shall indicate the proof. Suppose that $s = \lim s_n$ and let ε be a standard positive number. We know that there is an $n_0 \in N$ such that the L_R-statement

$$\forall x(N(x) \wedge x > n_0 \to |s - s_x| < \varepsilon)$$

is valid in R. By the extension principle it is valid in ${}^*R$. But any infinite ω is larger that n_0. Hence $|s - s_\omega| < \varepsilon$. Since this is valid for all standard ε, it follows that $s - s_\omega$ is infinitesimal.

Conversely, assume that $s - s_\omega$ is infinitesimal for all

$\omega \in {}^*N - N$. Let ε be standard positive and let $\upsilon \in {}^*N - N$. Obviously the statement

$$\forall x(\ N(x) \wedge\ x > \upsilon \ \rightarrow\ |s - s_x| < \varepsilon\ )$$

is true in ${}^*R$. But this is not a L_R-statement, as υ does not belong to R. However, the following L_R-sentence is true in ${}^*R$, and therefore in R,

$$\exists y(\ N(y)\ \wedge \forall x(\ N(x)\ \wedge\ x > y \rightarrow |s - s_x| < \varepsilon)).$$

Interpreting this statement in R, it follows at once that $\lim s_n = s$ in the usual sense.

We see that the proof consists of a shifting back and forth between R and ${}^*R$, using all the time the general extension principle.

Note that we have here identified a standard condition with a non-standard one. If we wanted to be exclusively non-standard, we could have used half of the theorem as a definition. This also applies to the following results. But our prime purpose is to explain how the infinitesimals give a correct foundation for analysis, hence we are primarily interested in characterization results.

If $a,b \in {}^*R$, we write $a \simeq b$ if $a-b$ is infinitesimal. $\simeq$ is an equivalence relation.- From the theorem above the following version of the Cauchy convergence criterion is hardly surprising:

THEOREM. <u>A standard sequence $\langle s_n \mid n \in N\rangle$ converges in the usual sense, if and only if $s_\lambda \simeq s_\mu$ for all infinitely large λ and μ.</u>

Our next task is to give the non-standard characterization of continuity. By a standard interval I in ${}^*R$ we understand the extension to ${}^*R$ of an interval in R. If I is given in R by the numbers a and b, the standard extension consists of all $x \in {}^*R$

such that $a < x < b$. The following result characterizes continuity.

THEOREM. A standard function $f(x)$ is continuous in the standard interval I, if and only if $f(x_0 + \eta) \simeq f(x_0)$ for all standard $x_0 \in I$ and all infinitesimals η such that $x_0 + \eta \in I$.

A characterization of uniform continuity is obtained by requiring the condition to be satisfied by all, and not only standard x_0 in I.

For the derivative we have the following result:

THEOREM. The standard number a is the derivative of $f(x)$ at $x = x_0$, if and only if for all infinitesimals η, $\eta \neq 0$,

$$\frac{f(x + \eta) - f(x)}{\eta} \simeq a.$$

Or, otherwise expressed: For all x in the domain of definition of $f(x)$, such that $dx = x - x_0$ is infinitely small, but not equal to 0, the ratio $\frac{df}{dx}$, where $df = f(x) - f(x_0)$, is infinitely close to $f'(x)$.

Ordinary mathematicians cannot say this. Leibniz tried, but with small success. Being non-standard mathematicians we may express ourselves in this way with a clean conscience.

It remains to characterize the integral in a non-standard way. Let $f(x)$ be a standard function which is continuous in the interval $a \leq x \leq b$, where a and b are standard. By a fine partition of $[a,b]$ we understand a sequence $x_0, x_1, \ldots, x_\omega$, where $\omega \in {}^*N - N$, $x_0 = a < x_1 < \ldots < x_\omega = b$, and $x_j - x_{j-1}$ is infinitesimal for $j = 1,2,\ldots,\omega$. There are (internal) fine partitions, e.g. let

$$x_j = a + \frac{b - a}{\omega} \cdot j$$

Let π be an internal fine partition and let ξ be an internal sequence $\xi_1, \xi_2, \ldots, \xi_\omega$ such that $x_j \leq \xi_{j+1} \leq x_{j+1}$, $j = 0,1,\ldots,\omega - 1$. We define

$$S(\pi,\xi) = \sum_{j=1}^{\omega} f(\xi_j) \cdot (x_j - x_{j-1}).$$

Here $S(\pi,\xi)$ has a meaning in $^*R$ as an extension of a well-defined functional in R. (This is precise enough, but the construction transcend our first-order logical frame-work. To give a complete explanation one ought to use a higher order language).

Note that seen from the inside of $^*R$ the infinite sum $S(\pi,\xi)$ is a "finite" sum. It seem that it is this fact, that certain operation in $^*R$ which from the outside are "infinite",nevertheless has a certain "finite" character, which makes the non-standard technique rather powerful from a heuristic point of view.

THEOREM. For every (internal) fine partion π and sequence ξ of the type considered above,

$$S(\pi,\xi) \simeq \int_a^b f(x)\,dx,$$

where the integral is taken in its standard sense.

To every element $a \in M_f$ there corresponds a uniquely determined number $a_0 \in R$ such that $a \simeq a_0$. We call a_0 the standard part of a, $a_0 = st(a)$. The integral is obtained by first taking a partition of the interval $[a,b]$ into infinitely many infinitesimal subintervals, and then the approximating sum. This sum is not necessarily equal to the integral, but its standard part gives the correct value.

We can see the importance of extending R to $^*R$. Within R "infinitely small intervals" are nothing but intervals of lenght 0, and any kind of approximating sum will also be 0. Within $^*R$ the infinitesimal intervals are distinct from 0 . And since all arithmetic operations extends from R to $^*R$, we have well-defined entities

$f(\xi_j) \cdot (x_j - x_{j-1})$. And since we can do finite sums in R, we can also do "finite" sums in $^*R$, hence the sum $\sum_{j=1}^{\omega} f(\xi_j)(x_j - x_{j-1})$ is well-defined as an element of $^*R$.

We shall not pursue the non-standard theory further. The reader is referred to a long series of works by A. Robeinson (of which we mention only a few in the bibliography). Robinson has not only reconstructed analysis in a non-standard way, he has also obtained many new results. Perhaps we ought to mention one, his solution of the so-called "invariant supspace" problem of Halmos and Smith. By use of non-standard techniques, in particular of a "finite" extension of the usual spectral theory of matrices, Robinson has shown that if T is a bounded linear operator of a Hilbert space H, such that a polynomial of T is compact, then T leaves invariant at least one closed linear subspace of H, different from H and {0}.-

But, perhaps the most interesting aspect of the theory is the fact that it gives a precise version of Leibniz's ideas. Leibniz was of the opinion that the infinitesimals and their calculus could be justified through the introduction of a kind of <u>ideal elements</u> which could be either infinitely small or infinitely large in comparision with the real numbers, but which should have the same properties as the reals. But Leibniz and his successors, in particular de l'Horpital, were not able to give a consistent development of the analysis based on these ideas. The introduction of $^*R$ accomplishes this, $^*R$ is an elementary extension of R, possesses, therefore, the same arithmetic properties. And the theory escapes the difficulties associated with Leibniz's attempt by carefully distinguishing between what "is equal" and what "is infinitely close to". For infinitesimal dx and corresponding dy, we have that $f'(x) \simeq \frac{dy}{dx}$, so that $f'(x) = st(\frac{dy}{dx})$, whereas Leibniz wanted to have $f'(x) = \frac{dy}{dx}$,- which is too much to ask for.

3. ON THE RELATIONSHIP BETWEEN STANDARD AND NON-STANDARD METHODS

Since non-standard models may be though of as ultraproducts, and the ultraproduct construction is well within the boundaries of ordinary mathematics, there is a sense on which non-standard methods are reducible to standard ones. However, this reduction is not very informative, and it does not throw much light on the relationship between methods in classical real analysis and non-standard methods.

This remark also applies to a reduction given by G. KREISEL [2] . He has shown by a metamathematical argument that given certain (reasonable) axiomatizations of non-standard and standard analysis, the former is a conservative extension of the latter.

In a note the present author has published on standard versus non-standard topology [1] , certain concepts of non-standard topology were reduced to the ordinary filter-theoretic ones. It seems that a further analysis of this reduction permits a better understanding of the relationship between ordinary and non-standard methods.

As a starting point let us consider the non-standard characterization of convergence of sequences. A sequence $\langle s_n \mid n \in N \rangle$ can be considered as a map $s: N \to R$, and as such it has an extension to a map ${}^*s: {}^*N \to {}^*R$. This led to the characterization that $\lim s_n = s$ iff $s \simeq s_\omega$ for all $\omega \in {}^*N - N$.

But s has also another extension. If the sequence $\langle s_n \mid n \in N \rangle$ is convergent, then the map $s: N \to R$ is bounded. And $s \in C^*(N)$, if N is given the discrete topology. s has then an extension, s^β, to βN, the Stone-Čech compactification of N. Let $\gamma \in \beta N - N$, it is obvious that $\lim s_n = s^\beta(\gamma)$. The connection between the two extension is then

$$s^{\beta}(\gamma) = st({}^{*}s(\omega)),$$

for all $\gamma \in \beta N - N$ and $\omega \in {}^{*}N - N$.

This relationship can be generalized. In order to do this we have to give some basic facts from the non-standard theory of topological spaces.

Note that if $X \subseteq M$, then X has a canonical extension ${}^{*}X$ to ${}^{*}M$. If e.g. ${}^{*}M = M^{I}/D$, then

$$f/_{D} \in {}^{*}X \text{ iff } \{i \in I \mid f(i) \in X\} \in D.$$

The injection of M into ${}^{*}M$ shows that $X \subseteq {}^{*}X$, but ${}^{*}X$ may in general be much larger than X. It may well happen that $\bigcap_{X \in F} X = \emptyset$ in M, but $\bigcap_{X \in F} {}^{*}X \neq \emptyset$ in ${}^{*}M$.

We shall now consider non-standard extensions ${}^{*}M$ of M which satisfies the following requirement:

(*) If F is a family of subsets of M with the finite intersection property, then there exist a point $a \in {}^{*}M$, such that $a \in {}^{*}X$ for all $X \in F$.

The existence of such non-standard models follows from the extended completeness theorem. It is also possible to give direct construction in terms of ultrapowers.

Let M be a topological space and ${}^{*}M$ a non-standard extension which satisfies (*). The points of M are called the standard points of ${}^{*}M$, and the canonical extension of the open sets of M are called the standard open sets of ${}^{*}M$. For every standard $a \in {}^{*}M$, define the monad of a, $\mu(a)$, as the intersection of all standard open sets which contains a as element. If $M = R$ and $b \in {}^*R$, then $a \in \mu(b)$ iff $a \simeq b$.

We have now the following connection between ultrafilters on M and points in ${}^*M$.

Given $b \in {}^*M$, we define a ultrafilter F_b on M by

$$X \in F_b \quad \text{iff} \quad b \in {}^*X,$$

i.e. F_b is the "trace" on M of the principal filter generated by the point b.

Conversely, given an ultrafilter F_λ on M, we can find a $b_\lambda \in {}^*M$ such that $b_\lambda \in {}^*X$ for all $X \in F_\lambda$. This follows immediately form the property $(*)$.

And $F_\lambda = F_{b_\lambda}$. (But different point in ${}^*M$ may map onto the same filter in M.) This implies that if a is a standard point of ${}^*M$, then $b \in \mu(a)$ iff F_b is finer that the neighborhoodfilter $V(a)$ of a in M. In our previous note [1] we showed how this reduced the non-standard characterization of compactness to the known one in terms of ultrafilters. And it is this very same correspondence which "explains" the non-standard characterization of continuity.

Recall that $f(x)$ is continuous at x_0, iff $f(x_0 + \eta) \simeq f(x_0)$ for all infinitesimals η.

An infinitesimal η in R defines an ultrafilter which converges to 0. $x_0 + \eta$ will represent a translation of this filter and define one which converges to x_0. Known properties of ultrafilters then implies that $f(x_0 + \eta)$ defines an ultrafilter which is the image by f of the filter determined by $x_0 + \eta$. And the condition $f(x_0 + \eta) \simeq f(x_0)$ says that $f(x_0 + \eta) \in \mu\,(f(x_0))$,i.e. the filter defined by $f(x_0 + \eta)$ converges to $f(x_0)$,- and we have obtained a translation of the non-standard characterization to the usual definition of continuity.

This correspondence between ultrafilters and non-standard points can be generalized in the following way.

Let M be a completely regular topological space, we have the following commutative diagram:

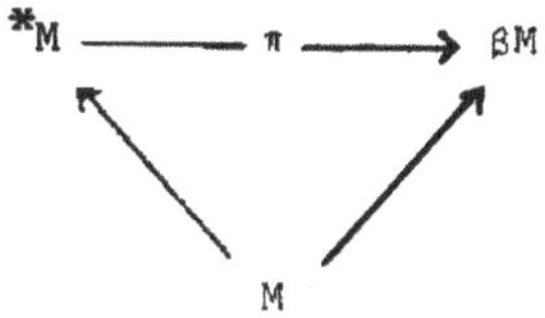

where the map π is onto. The map π is defined in the following way. Every $\alpha \in {}^*M$ induces a prime z-filter on M viz. $F_\alpha \cap Z(M)$, where $Z(M)$ is the collection of zero-sets on M. This z-filter can be extended to a unique ultra z-filter, and βM is the collection of all ultra z-filters on M. (This correspondence, which was implicit in the authors note [1] , has also been explicitly given by A.Robinson).

Let $f \in {}^*C(M)$. We have two extensions ${}^*f\colon {}^*M \to {}^*R$ and $f^\beta\colon \beta M \to R$. The following identity holds for all $\alpha \in {}^*M$:

$$(**) \qquad f^\beta(\pi(\alpha)) = st({}^*f(\alpha)).$$

The proof consists in showing that both $st({}^*f(\alpha))$ and $f^\beta(\pi(\alpha))$ are equal to $\lim_{F_\alpha} f$, where F_α is the ultrafilter induced on M by α.

Since f is bounded, $st({}^*f(\alpha)) = r$ for some $r \in R$. This means that ${}^*f(\alpha)$ belongs to the monad of r, i.e. $F_{f(\alpha)}$ is finer than the neighborhoodfilter $V(r)$. We have to show that $f(F_\alpha)$ converges to r. Since $F_{f(\alpha)}$ converges to r, we need only show that given any $X_0 \in F_{f(\alpha)}$, we can find some $X_1 \in f(F_\alpha)$ such that $X_1 \subseteq X_0$. Thus let X_o be given, and let $Y = f^{-1}(X_o)$. Since ${}^*f(\alpha) \in {}^*X_o$, we have that $\alpha \in ({}^*f)^{-1}({}^*X_o) = {}^*(f^{-1}(X_o)) = {}^*Y$, i.e. $Y \in F_\alpha$. We may choose $X_1 = f(Y) = X_o$.

Recall that $\pi(\alpha)$ was defined as the unique ultra z-filter which extends the prime z-filter $F_\alpha \cap Z(M)$. It is well-known that the filter $\pi(\alpha)$ and the filter F_α converges to the same point in βM, viz. the point $\pi(\alpha)$. Hence by the continuity of the extension f^β, $f^\beta(\pi(\alpha)) = \lim_{F_\alpha} f^\beta = \lim_{F_\alpha} f$.

It is the commutative diagram above and the result (**) which effects a partial reduction of the non-standard theory to the standard one. We treated the non-standard characterization of the limit of a sequence in the introduction to this section. A similar "explanation" of the non-standard definition of the integral is possible. In this case we consider the space T of all partitions $\langle\pi,\xi\rangle$, where π is a partion $x_o = a, x_1,\ldots,x_n = b$ of the interval $[a,b]$ and ξ is a sequence of numbers $\xi_1,\ldots,\xi_n$ where $x_{i-1} \leq \xi_i \leq x_i$, $i = 1,\ldots,n$. On the (discrete) space T we define a map $I_f = T \to R$ by

$$I_f(\pi,\xi) = \sum_{i=1}^{n} f(\xi_i) \cdot (x_i - x_{i-1}).$$

The identity (**) applied to the extensions ${}^*I_f$ and I_f^β gives the non-standard characterization of the integral. (Warning: $I_f^\beta(\lambda)$ is not equal to the integral for all $\lambda \in \beta T - T$.) Note, however, that whereas an $\alpha \in {}^*T$ admits a representation as an infinite partition, and hence ${}^*I_f(\alpha)$ as an infinite sum $\sum_{i=1}^{\omega} f(\xi_i) \cdot (x_i - x_{i-1})$, we have no similar representation for points in $\beta T - T$.

The commutative diagram and the property (**) extend to the

case where M is a uniform space, βM is replaced by the completion of M, and f is assumed to be uniformly continuous.

One may also use $^*M$, if M is a uniform space, to construct the completion of M. An element $b \in {}^*M$ is called <u>bounded</u> if it belongs to the intersection in $^*M$ of some family $\mathfrak{X}$ of closed sets in M which has the finite intersection property and which contains arbitrarily small sets. It is easy to see that the set of bounded elements in $^*M$ "modulo monads" is the completion of M in the given uniformity. And it is also easy to see that this is nothing but the lifting from βM to $^*M$ of the well-known construction of the completion using the Stone-Čech compactification.

We claimed that the reduction in a certain sense is complete if M is a topological or uniform space. If M also carries an algebraic structure, the situation is much more involved. If M is a topological group, then βM is not in general a group, but βM is "nice" with respect to topological properties. $^*M$, as an elementary extension of M, is certainly a group, but it is not a topological group (in the weakest topology on $^*M$ which makes π continuous). Thus none of the extensions $^*M$ and βM can replace the other, but by simultaneously using both extensions, it is possible to obtain many interesting results. (See e.g. A. Robinson's paper [4] on non-standard arithmetic).

REMARK. Using the bounded elements also seems to be the proper setting for discussing the completion of algebraic structures. In this case M is both an algebraic structure and a uniform space, and it is ordinarily assumed that the algebraic operations are continuous in the topology deduced from the uniformity. (But the algebraic operations are not necessarily uniformly continuous). A necessary condition for the completion of M to exists, is that the bounded elements of $^*M$ be closed under the extended algebraic operations of M. This condition is not in general sufficient. One must also require a certain continuity property of the extended operations, which leads to a necessary and sufficient condition.

"Infinitely close" entities are equal in βM. It is in $^{*}M$ that we can write the integral as an infinite sum. It is $^{*}M$ which gives the correct formalism in the spirit of Leibniz.

One may argue how interesting this formalism is. But what is beyond doubt, is the fact that the ultraproduct is an extremely powerful construction. So even if the logician first defined it, it ought to become a standard tool for the ordinary mathematician.

REFERENCES

[1]. J.E. FENSTAD, A note on "standard" versus "non-standard" topology. Indag.Math. 29(1967),pp. 378-380.

[2]. G. KREISEL, Axiomatizations of non-standard analysis which are conservative extensions of formal systems for classical standard analysis.
Forthcoming in Synposium on Applications of Model Theory to Analysis and Algebra.

[3]. A. ROBINSON, Non-standard analysis.
North-Holland Publ. Comp., Amsterdam 1966.

[4]. A. ROBINSON, Non-standard arithmetic.
Bull. AMS., 73(1967),pp. 818-843.

On Factorization of Manifolds

Peter Hilton

1. Introduction

Examples were constructed in [1] of the following phenomenon. We consider based homotopy types of polyhedra. There is a natural addition (union with base points identified) giving the class of homotopy types the structure of a commutative semigroup with 0. Then this semigroup is not a cancellation semigroup. That is, one finds (many!) examples of polyhedra P_1, P_2, A such that

$$P_1 + A \simeq P_2 + A \quad , \quad P_1 \not\simeq P_2 \tag{1}$$

The question naturally arises whether the construction of such examples (1) may be dualized (in the sense of Eckmann-Hilton-Fuks) to yield examples of the dual phenomenon; that is, to construct examples of polyhedra Q_1, Q_2, B such that

$$Q_1 \times B \simeq Q_2 \times B, \quad Q_1 \not\simeq Q_2. \tag{2}$$

Of course, the problem of cancellation for topological products is classical. It is known that cancellation of topological types is not valid. Also since cancellation of factors is not valid for the class of finitely presented groups, it follows that cancellation of factors is not valid for non-simply-connected homotopy types (we naturally assume the polyhedra have finitely many cells in each dimension, otherwise cancellation is plainly invalid).

Now it turns out that the construction yielding examples of (1) may be dualized to yield examples of (2) but it will yield infinite-dimensional examples. For we construct P_1, P_2 in (1) as spaces with two non-vanishing homology groups so that Q_1, Q_2 in (2) would be spaces with two non-vanishing homotopy groups - and B would be an Eilenberg-MacLane space. Moreover, Q_1 and Q_2 would be distinguished by their k-invariants. Such examples, while moderately interesting, do not satisfy one's appetite for examples of (2) where Q_1, Q_2, B are compact polyhedra (and, preferably, 1-connected).

Such examples may indeed be constructed by replacing two-stage Postnikov systems by principal bundles. This modification of the procedure leads to a considerable complication of the argument but produces a very strong result. For the examples we obtain have the following properties. We construct smooth, closed manifolds M_1, M_2 which are total spaces of principal G-bundles for some Lie group G, and are such that

(3) $$M_1 \times G = M_2 \times G, \quad M_1 \not\cong M_2.$$

Here the equality sign means that the product-manifolds are diffeomorphic. One such example was given in [2], but the crude techniques employed there did not permit us to exploit the construction to its fullest extent. In particular, the **crucial** theorem (Theorem 3 below) was missing.

A particular example of (3) is of independent interest. We may take $M_1 = Sp(2)$, $G = S^3$. Then there is a closed, smooth 10-manifold M which is a principal S^3-bundle over S^7 and satisfies

(4) $$M \times S^3 = Sp(2) \times S^3, \quad M \not\simeq Sp(2).$$

Thus M is a "new" H-manifold, being a retract of a Lie group.

The work reported on here is joint work with J. Roitberg and will be published in detail elsewhere.

2. <u>The construction</u>

We first recall the construction of examples of (1) in its simplest form. Let $\alpha \in \pi_{m-1}(S^n)$ be of order k, let ℓ be prime to k and let $\beta = \ell\alpha$. Let C_α, C_β be the mapping cones of α, β (thus C_α, C_β are really homotopy types, but the abuse of terminology is justified!). We thus have (see [1])

<u>Theorem 1</u>. $$C_\alpha \simeq C_\beta \Longleftrightarrow (\pm 1) \circ \beta = \alpha \circ (\pm 1).$$

<u>Thus, if α is a suspension element</u>, $C_\alpha \simeq C_\beta \Longleftrightarrow \beta = \pm\alpha$.

<u>Theorem 2</u>. $$C_\alpha + S^m \simeq C_\beta + S^m$$

<u>Corollary 1</u>. <u>Let</u> α <u>be a suspension element of order</u> $p \neq 2,3$, <u>let</u> ℓ <u>be prime to</u> p, $\ell \not\equiv \pm 1$ <u>mod</u> p, <u>and</u> $\beta = l\alpha$. <u>Then</u>

$$C_\alpha + S^m \simeq C_\beta + S^m, \quad C_\alpha \not\simeq C_\beta.$$

We also recall the proof of Theorem 2. We consider the diagrams, for any $\alpha, \beta \in \pi_{m-1}(S^n)$,

(5) $$\begin{array}{ccccc} S^{m-1} & \xrightarrow{\beta} & S^n & \xrightarrow{i_\alpha} & C_\alpha \\ & & \downarrow{\scriptstyle i_\beta} & & \downarrow \\ & & C_\beta & \longrightarrow & C_{\alpha\beta} \end{array}$$

where i_α, i_β embed S^n in the mapping cones C_α, C_β and we have constructed the cofibre-product (= push-out) of i_α and i_β. Then $C_{\alpha\beta} = C_{\beta\alpha}$ and

$$C_{\alpha\beta} \simeq C_\alpha + S^m \qquad \text{if } i_\alpha \circ \beta = 0.$$

Now if $\beta = \lambda\alpha$, then

$$i_\alpha \circ \beta = i_\alpha \circ \lambda\alpha = \lambda(i_\alpha \circ \alpha) = 0 \quad \text{since } i_\alpha \circ \alpha = 0.$$

But since λ is prime to k we have $\alpha = \bar{\lambda}\beta$ for some $\bar{\lambda}$ so that also $i_\beta \circ \alpha = 0$ and

$$C_{\beta\alpha} \simeq C_\beta + S^m.$$

We could, as we have said, dualize this construction, replacing S^{m-1}, S^n by Eilenberg-MacLane spaces. Thus let $\alpha \in H^{m+1}(Z,n;Z)$ be of order k, λ prime to k, $\beta = \lambda\alpha$; this implies $m > n$. If E_α is the <u>mapping path-space</u> of α, that is, if $p_\alpha : E_\alpha \to K(Z,n)$ is induced by α, we have the theorems, dual to those above:

<u>Theorem 1'</u>. $\quad E_\alpha \simeq E_\beta \Longleftrightarrow (\pm 1) \circ \beta = \alpha \circ (\pm 1)$.

<u>Thus, if α is a suspension element</u>, $E_\alpha \simeq E_\beta \Longleftrightarrow \beta = \pm\alpha$.

<u>Theorem 2'</u>. $\quad E_\alpha \times K(Z,m) \simeq E_\beta \times K(Z,m)$.

<u>Remark</u>. Theorems 1 and 1' hold for <u>any</u> elements α, β in the appropriate groups.

The proof of Theorem 2' utilizes a diagram dual to (5),

$$(5') \qquad \begin{array}{ccccc} E_{\alpha\beta} & \longrightarrow & E_\beta & & \\ \downarrow & & \downarrow{\scriptstyle p_\beta} & & \\ E_\alpha & \xrightarrow{p_\alpha} & K(Z,n) & \xrightarrow{\beta} & K(Z,m+1) \end{array}$$

where we have constructed the fibre-product (= pull-back) of p_α and p_β.

We now modify this dual construction, replacing $K(Z,m+1)$ by the classifying space, B_G, of a Lie group G. In fact we will be exclusively concerned from here on with the case $G = S^3$. Also we will consider principal S^3-bundles over S^n. Such bundles are classified (up to bundle-equivalence) by elements of $\pi_{n-1}(S^3)$. Let $\alpha \in \pi_{n-1}(S^3)$ correspond to $\alpha_0 \in \pi_n(B_{S^3})$ under the canonical (adjoint) isomorphism. We will adopt the notation

$$\begin{array}{ccc} S^3 \overset{i_\alpha}{\subseteq} E_\alpha & & \\ \downarrow{\scriptstyle p_\alpha} & & \\ S^n & \xrightarrow{\alpha_0} & B, \end{array}$$

$B = B_{S^3}$; thus we confuse maps and homotopy classes as before!

<u>Proposition 1</u> (James-Whitehead [3]) E_α <u>has a CW-structure</u>

$$E_\alpha = (S^3 \cup_\alpha e^n) \cup e^{n+3}$$

<u>Theorems 1"</u> $E_\alpha \simeq E_\beta \Longleftrightarrow \beta = \pm\alpha$.

<u>Proof</u>. If $\beta = \pm\alpha$, then, in fact, $E_\alpha = E_\beta$. Conversely suppose $E_\alpha \simeq E_\beta$. We distinguish three cases:

(i) $n \leq 3$ - trivially $\beta = \pm\alpha$ because $\pi_{n-1}(S^3) = 0$.

(ii) $n = 4$ - then α, β are integers, $H_3(E_\alpha) = Z_{|\alpha|}$, $H_3(E_\beta) = Z_{|\beta|}$, so that $|\alpha| = |\beta|$.

(iii) $n \geq 5$ - then we may assume an equivalence $h: E_\alpha \simeq E_\beta$ such that $h(S^3) \subseteq S^3$ (see Proposition 1).

Then h is of degree ± 1 on S^3 and, by the exact homotopy sequence, $h_*: \pi_n(E_\alpha, S^3) \cong \pi_n(E_\beta, S^3)$. These groups are, of course, cyclic infinite and we have a commutative square

$$\begin{array}{ccc} \pi_n(E_\alpha, S^3) & \overset{\pm 1}{\cong} & \pi_n(E_\beta, S^3) \\ \downarrow \partial & & \downarrow \partial \\ \pi_{n-1}(S^3) & \overset{\pm 1}{\cong} & \pi_{n-1}(S^3) \end{array}$$

Starting with a generator of $\pi_n(E_\alpha, S^3)$ and going one way round the square gives $\pm\alpha$, the other way round gives $\pm\beta$.

To obtain a result corresponding to Theorems 2,2' we must consider the diagram

$$(5'') \qquad \begin{array}{ccccc} E_{\alpha\beta} & \longrightarrow & E_\beta & & \\ \downarrow & & \downarrow p_\beta & & \\ E_\alpha & \xrightarrow{p_\alpha} & S^n & \xrightarrow{\beta_o} & B \end{array}$$

Then

(6) $$E_{\alpha\beta} = E_{\beta\alpha} \quad \text{and} \quad E_{\alpha\beta} = E_{\alpha} \times S^3 \quad \text{if } \beta_0 \circ p_\alpha = 0.$$

However, the argument proving Theorem 2 does not hold here, since, even if $\beta_o = \ell\alpha_o$ we cannot write $\beta_0 \circ p_\alpha = \ell(\alpha_o \circ p_\alpha)$, because $\Pi(E_\alpha, B)$ has no group structure. Indeed, it turns out that it is not always true that $\ell\alpha_o \circ p_\alpha = 0$, although, of course, $\alpha_0 \circ p_\alpha = 0$. However, we may prove the following theorem. As before we assume α to be of order k.

<u>Theorem 3</u>. <u>Let</u> $\ell \equiv \ell'$ mod k <u>where</u>

$$\frac{\ell'(\ell'-1)}{2}\,\omega \circ \Sigma^3\alpha = 0,$$

$\omega \in \pi_6(S^3)$ <u>being the generator (which measures the non-commutativity of</u> S^3). <u>Then if</u> $\beta = \ell\alpha$,

$$\beta_o \circ p_\alpha = 0 \ \underline{\text{so that}}\ E_{\alpha\beta} = E_\alpha \times S^3.$$

This is proved by analysing the Puppe sequence of i_α,

(6) $$S^3 \xrightarrow{i_\alpha} E_\alpha \xrightarrow{j_\alpha} S^n + S^{n+3} \xrightarrow{u_\alpha} S^4 .$$

One shows that u_α has components $u_\alpha = \langle \Sigma\alpha, \gamma \circ \Sigma^4\alpha \rangle$, where $\gamma \in \pi_7(S^4)$ is the Hopf map. Also p_α factorizes as $q_\alpha j_\alpha$ and

$$\alpha_0 \circ q_\alpha = e \circ u_\alpha = \langle \alpha_0, o \rangle$$

where $e \in \pi_4(B)$ is adjoint to $1 \in \pi_3(S^3)$. Then

$$\ell'\alpha_o \circ q_\alpha = \langle \ell'\alpha_o, 0 \rangle$$

$$\ell' e \circ u_\alpha = \langle \ell'\alpha_o, \ell' e \circ \gamma \circ \Sigma^4\alpha \rangle$$

$$= \langle \ell'\alpha_o, \frac{\ell'(\ell'-1)}{2}[e,e] \circ \Sigma^4\alpha \rangle.$$

Since the Whitehead product $[e,e]$ is adjoint to ω, the theorem readily follows.

<u>Theorem 4</u>. <u>Let</u> α <u>be of order</u> k, $k_0 = \gcd(k,24)$, ℓ <u>prime to</u> k, $\ell \equiv 1$ <u>mod</u> k_o, $\beta = \ell\alpha$. <u>Then</u>

$$E_\alpha \times S^3 = E_\beta \times S^3.$$

Proof Now $12\,\omega = 0$. Let us choose $\ell' \equiv \ell \bmod k$, $\ell' \equiv 1 \bmod 24$; this is possible since $\ell \equiv 1 \bmod k_0$. Then $\frac{\ell'(\ell'-1)}{2}\,\omega \circ \Sigma^3\alpha = 0$, so that $E_{\alpha\beta} = E_\alpha \times S^3$. But the residues prime to k and congruent to 1 mod k_0 form a group. Thus we may find $\bar{\ell}'$ with $\alpha = \bar{\ell}\beta$ and $\bar{\ell}$ prime to k, $\bar{\ell} \equiv 1 \bmod k_0$. This means that we may exchange the roles of α and β and deduce $E_{\beta\alpha} = E_\beta \times S^3$.

<u>Corollary 1"</u>. <u>Let</u> α <u>be of order</u> $p \neq 2,3$, <u>let</u> ℓ <u>be prime to</u> p, $\ell \not\equiv \pm 1$ <u>mod</u> p, <u>and</u> $\beta = \ell\alpha$. Then

$$E_\alpha \times S^3 = E_\beta \times S^3 \quad , \quad E_\alpha \not\simeq E_\beta .$$

<u>Remark</u>. We have $E_\alpha = C_\alpha \cup e^{n+3}$, $E_\beta = C_\beta \cup e^{n+3}$. We may also show that, under the same hypotheses as those of Corollary 1",

$$C_\alpha + S^3 \simeq C_\beta + S^3 \ , \ C_\alpha \not\simeq C_\beta .$$

Moreover, the homotopy equivalence $C_\alpha + S^3 \simeq C_\beta + S^3$ cannot, in general, extend to a homotopy equivalence $C_\alpha \times S^3 \simeq C_\beta \times S^3$, although $E_\alpha \times S^3 = E_\beta \times S^3$. For, if it could, then we could pass to quotients and obtain a homotopy equivalence $\Sigma^3 C_\alpha \simeq \Sigma^3 C_\beta$ or $C_{\Sigma^3\alpha} \simeq C_{\Sigma^3\beta}$. But if $\Sigma^3\alpha \neq 0$ (for example, if $\alpha \in \pi_{2p}(S^3)$) then $C_{\Sigma^3\alpha} \not\simeq C_{\Sigma^3\beta}$ by Theorem 1.

Corollary 2 (of Theorem 3). Let $M = E_{7\omega}$. Then $M \not\approx Sp(2)$ but

$$M \times S^3 = Sp(2) \times S^3.$$

For $\pi_q(S^3) = Z_3$. Thus if $\alpha = \omega$ (so that $k = 12$) and $\ell = 7$, we have $\frac{\ell(\ell-1)}{2}\,\omega \cdot \Sigma^3\omega = 0$, so that $E_{\alpha\beta} = E_\omega \times S^3$, where $\beta = 7\omega$. But then $\alpha = 7\beta$, so the situation is symmetrical in α and β and $E_{\beta\alpha} = E_{7\omega} \times S^3$. Since $E_\omega = Sp(2)$, the corollary follows from Theorem 1".

Bibliography

1. Peter Hilton, On the Grothendieck group of compact polyhedra, Fund. Math 61(1967), 199-214.

2. Peter Hilton and Joseph Roitberg, Note on principal S^3-bundles, Bull. A.M.S. (1968).

3. I.M. James and J.H.C. Whitehead, The homotopy theory of sphere-bundles over spheres I, Proc. L.M.S. 4(1954), 196-218.

Courant Institute of Mathematical Sciences, New York University
Cornell University

HOMEOMORPHISMS WITH A GIVEN DILATATION

Olli Lehto, University of Helsinki

1. Classical Beltrami equation. There are many natural ways to arrive at the Beltrami differential equation

(1)
$$w_{\bar{z}} = \mu w_z ,$$

where the complex-valued function μ is of modulus < 1. For instance, if w is a sense-preserving diffeomorphism of a plane domain A, then the Jacobian $|w_z|^2 - |w_{\bar{z}}|^2$ of w is positive throughout A. Hence, every such mapping w satisfies an equation (1) with $|\mu(z)| < 1$.

The coefficient μ, the complex dilatation of w, allows a simple geometric interpretation: $\mu(z)$ describes the infinitesimal ellipse at z which w maps on an infinitesimal circle. The absolute value $|\mu(z)|$ determines the eccentricity, which tends to 1 with $|\mu(z)| \to 1$, and $\arg \mu(z)$ gives the direction of the ellipse. The vanishing of μ in A is equivalent to w being conformal in A. It follows that complex dilatation determines a diffeomorphism up to a conformal transformation. This gives rise to the question, what functions μ can be complex dilatations for diffeomorphisms.

The famous classical problem of making a sufficiently smooth orientable surface S in the euclidean 3-space into a Riemann surface also leads to equation (1). A differentiable S carries a natural metric which, expressed in terms of local coordinates, is of the form

(2)
$$ds = \lambda |dz + \mu d\bar{z}| .$$

Here μ is complex-valued and $|\mu(z)| < 1$. In the case $\mu = 0$, the coordinates are called isothermal. It is well known that a system of isothermal coordinates defines a conformal structure for S. If a coordinate transformation $z \to w$ is performed in the plane, a simple computation shows that the new coordinates w are isothermal if w is a locally injective solution of (1) where μ is the function in (2).

It was in connection with this problem that Gauss in 1822 proved the existence of locally injective solutions of (1), provided that μ is real analytic. Owing to the above-mentioned uniqueness property, these local solutions w define a conformal structure W for A. By the uniformization theorem, there exists a conformal map f from the Riemann surface A with the conformal structure W into the plane. Locally $f = \phi \circ w$, where ϕ is conformal in the ordinary sense. Hence, f is an injective solution of (1) in A.

The result of Gauss was generalized by Lichtenstein and Korn in 1914-16; they proved that the result remains valid if μ is Hölder-continuous. The case of an arbitrary continuous μ turned out to be more difficult. Using a different terminology, Lavrentjev in 1935 established a mapping theorem by which a homeomorphic solution of (1) exists also if μ is just continuous. The result was translated into the language used here by Bojarski in 1957. However, a new phenomenon was encountered: In contrast to the Hölder-continuous case, the solution need not be differentiable at every point of A. The concept of a solution must therefore be redefined.

This was done in 1938 by Morrey. He assumed that μ is only measurable with respect to the 2-dimensional Lebesgue measure. The additional assumption that $|\mu|$ be bounded away from 1 in every compact subset of A is needed; in the case of a continuous μ this condition is of course automatically fulfilled. A function w is called a solution of (1) if its generalized L^2-derivatives satisfy (1). This means that w is absolutely continuous on lines in A, it is differentiable a.e., and (1) holds a.e. It follows from Morrey's results that with these assumptions on μ and this definition of solution, equation (1) always possesses homeomorphic solutions w. They are locally quasiconformal mappings; one says that w is locally μ-quasiconformal. This existence theorem, which is fundamental in the theory of plane quasiconformal mappings, has been subsequently proved in many different ways. It is still true in this general case that the solutions are unique up to conformal mappings. For more details we refer to [2].

2. A.e. quasiconformal homeomorphisms. Let us now consider a slightly different existence problem, still closely related with finding homeomorphic solutions for (1), which seems to be important in some applications. In the following, A will be the whole extended plane. The set A-B is denoted by -B.

Let E be a compact set of two-dimensional measure zero and μ a measurable function in the plane such that $\sup\{|\mu(z)|\,|z\in F\}<1$ for every compact set $F\subset -E$. E plays the role of an exceptional set: it is possible that $|\mu(z)|\to 1$ as z approaches E. A function w is called a μ-homeomorphism if w is homeomorphic in the plane and locally μ-quasiconformal in -E.

Simple examples show that μ-homeomorphisms may fail to exist even if E consists of a single point. On the other hand, if $|\mu|$ is bounded away from 1, the above-mentioned existence theorem quarantees the existence of μ-homeomorphisms irrespective of the set E. We shall show that the same remains true under weaker restrictions on the growth of μ near E.

For a given set E, the existence of μ-homeomorphisms depends only on the behavior of μ in an arbitrary small neighborhood of E:

Theorem 1. Let w_0 be a μ_0-homeomorphism and $D_1,\ldots,D_n$ disjoint Jordan domains whose union covers E. Then a μ-homeomorphism w exists where μ equals μ_0 in $\cup D_i$ and can be given arbitrarily elsewhere.

Proof: We can obviously assume that every D_i is bounded by a rectifiable curve. Let $f_i:D_i\to w_0(D_i)$ be a quasiconformal mapping with boundary values w_0. By the above existence theorem, we can find a quasiconformal mapping g of the plane whose complex dilatation agrees a.e. with the complex dilatation of f_i in D_i and is equal to the given μ in the complement of $\cup D_i$. Then the mapping w, $w = g\circ f_i^{-1}\circ w_0$ in D_i, $w = g$ in $-\cup D_i$, is a μ-homeomorphism. It is clearly homeomorphic in the whole plane and locally μ-quasiconformal in $\cup D_i\cap E$ as well as in $-\cup\overline{D}$. Quasiconformality on the boundary of D_i follows from the fact that as a rectifiable curve the boundary is removable.

In addition to the existence problem, there are many other natural problems concerning μ-homeomorphisms. It is no longer always true that μ-homeomorphisms are unique up to conformal mappings of the plane, i.e. up to linear transformations. Even 0-homeomorphisms need not all be conformal if the exceptional set E is large. So the question arises, under what conditions μ-homeomorphisms are unique up to linear transformations.

We can also ask, when it is possible to extend every locally μ-quasiconformal homeomorphism of -E to a μ-homeomorphism. Or, we can ask, when it is true that a set of zero area maps on a set of zero area under a μ-homeomorphism. In this paper, we shall restrict ourselves to the existence problem.

3. Modulus of a ring domain. The proof of the existence theorem is based on estimates of the conformal modulus M of doubly connected domains. All inequalities below can be derived in a unified manner by expressing the modulus as the extremal distance of the boundary components. In Lemmas 1 and 2 we use the spherical metric $(1 + |z|^2)^{-1}|dz|$; s stands for the distance in this metric.

L e m m a 1. Let A be a ring domain which separates the points a_1, b_1 from the points a_2, b_2. If $s(a_i, b_i) \geq \eta$, $i = 1,2$, then

$$M(A) \leq \frac{\pi^2}{2\eta^2} .$$

For the proof we refer to [2], p. 36.

L e m m a 2. Let A be a ring domain whose boundary components have a spherical diameter $\geq \delta$ and a spherical distance $\leq \varepsilon$, $\varepsilon < \delta$, from each other. Then

$$M(A) \leq \frac{\pi^2}{\log \frac{\tan \delta/2}{\tan \varepsilon/2}} .$$

([2], p. 36).

Set

$$\psi_\mu(z,r,\theta) = \frac{|1-e^{-2i\theta}\mu(z+re^{i\theta})|^2}{1-|\mu(z+re^{i\theta})|^2}, \quad \chi_\mu(z,r,\theta) = \frac{|1+e^{-2i\theta}\mu(z+re^{i\theta})|^2}{1-|\mu(z+re^{i\theta})|^2}.$$

For $z = 0$, we write $\psi_\mu(r,\theta)$, $\chi_\mu(r,\theta)$.

L e m m a 3. Let $A = \{z \mid r_1 < |z| < r_2\}$, E a closed set in A which is of σ-finite linear measure, and f a homeomorphism of a domain $D \supset \overline{A}$. which is locally μ-quasiconformal in $A-E$. Then

(3)
$$2\pi \int_{r_1}^{r_2} \frac{dr/r}{\int_0^{2\pi} \psi_\mu(r,\theta)d\theta} \leq M(f(A)) \leq \frac{2\pi}{\int_0^{2\pi} \frac{d\theta}{\int_{r_1}^{r_2} \chi_\mu(r,\theta)dr/r}}.$$

Proof: For a mapping f, quasiconformal in A, (3) was established by Reich and Walczak [5]. After suitable modifications, the same proof applies to the present situation since almost all radii from the origin and almost all circles centered at the origin meet E at not more than countably many points.

Denote

$$\psi_\mu^*(r) = \frac{1}{2\pi} \int_0^{2\pi} \psi_\mu(r,\theta)d\theta,$$

and similarly for χ_μ. Then it follows from (3) that

(4)
$$\int_{r_1}^{r_2} \frac{dr}{r\psi_\mu^*(r)} \leq M(f(A)) \leq \int_{r_1}^{r_2} \frac{\chi_\mu^*(r)dr}{r}.$$

The left side inequality is the same as in (3), and the right hand estimate follows from Schwarz's inequality.

The function

$$D = \frac{1 + |\mu|}{1 - |\mu|}$$

is called the dilatation quotient. Since $\psi_\mu \leq D$,

$$\text{(5)} \qquad M(f(A)) \geq \int_{r_1}^{r_2} \frac{dr}{rD^*(r)} .$$

Here again, as in what follows, * stands for the mean value over $(0,2\pi)$.

4. <u>Convergence of quasiconformal mappings</u>. In the proof of the existence theorem we shall make use of the following result on the limit of a convergent sequence of quasiconformal mappings.

L e m m a 4. Let f_n, $n = 1,2,\ldots$, $f_n(a_i) = a_i$, $i = 1,2,3$, be a sequence of quasiconformal mappings of the extended plane which converges uniformly towards a limit function f. If for every annulus A, the numbers $M(f_n(A))$ are bounded away from zero, then f is a homeomorphism.

Proof: We remark that if all mappings f_n are K-quasiconformal for a fixed K, then $M(f_n(A)) \geq M(A)/K > 0$. Also, if the mappings f_n are μ_n-quasiconformal with $|\mu_n| \leq |\mu|$, where $1/(1-|\mu|)$ is locally integrable, then using (5) we obtain a uniform positive lower bound for $M(f_n(A))$.

Since f is continuous, the lemma follows if we prove that f is one to one. Suppose that $f(z_1) = f(z_2) = w_0$, $z_1 \neq z_2$; we can assume that $w_0 \neq \infty$. Let A be an annulus such that one of its boundary curves contains z_1 and one of the points $a_i \neq w_0$. The other boundary circle contains z_2 and one of the remaining points $a_i \neq w_0$. Save for some exceptional locations of the points z_1, z_2, and a_i, such an annulus exists.

Let $\varepsilon > 0$; we choose ε so small that at least two of the points a_i lie outside the disc $D_\varepsilon = \{w | s(w,w_0) < \varepsilon\}$. Let n_0 be an integer such that $f_n(z_i) \varepsilon D_\varepsilon$, $i = 1,2$, $n \geq n_0$. For $n \geq n_0$ the ring domain $f_n(A)$ then has the following properties. The spherical diameters of its boundary components are bounded away from zero: both contain a point in D_ε and one of the points a_i outside D_ε. On the other hand, the spherical distance of the boundary components is $< 2\varepsilon$. Hence, by Lemma 2, $M(f_n(A)) \to 0$ as $n \to \infty$. This contradicts the hypothesis.

If the above annulus A cannot be constructed we replace z_1 by a suitable point z_1' such that $s(f(z_1), f(z_1')) < \varepsilon/2$ and derive a contradiction as above.

5. Sufficient existence condition. By virtue of the above Lemmas, we can now establish a sufficient condition for the existence of μ-homeomorphisms.

T h e o r e m 2. Let

$$\int_{r_1}^{r_2} \frac{dr}{r(1+\psi_\mu^*(z,r))}, \qquad r_2 > r_1,$$

be positive for every finite z and tend to ∞ as $r_1 \to 0$ or as $r_2 \to \infty$. Then μ-homeomorphisms exist.

Proof: Let us first remark that the above condition is certainly fulfilled if $|\mu|$ is bounded away from 1, because $\psi_\mu^*(z,r)$ is then uniformly bounded in z and r.

Let $\{G_n\}$ be an exhaustion of $-E$ by open sets G_n. Let μ_n be the function which equals μ in G_n and vanishes in $-G_n$, and f_n the μ_n-quasiconformal mapping of the plane normalized by the conditions $f_n(a_i) = a_i$, $i = 1,2,3$. We prove that the family $\{f_n\}$ is equicontinuous with respect to the spherical metric at every point of the extended plane.

Choose first an arbitrary finite point z_0 and then a number $\rho > 0$ so small that at least two of the points a_i, say a_1 and a_2, lie outside the disc $|z-z_0| \leq \rho$. Having fixed ρ we consider the annulus $A_\delta = \{z \mid \delta < |z-z_0| < \rho\}$. If z_1 is an arbitrary point such that $|z_1-z_0| < \delta$, then $f_n(A_\delta)$ separates the points $f_n(z_0)$, $f_n(z_1)$ from the points a_1, a_2. Therefore, by Lemma 1,

$$M(f_n(A_\delta)) \leq \frac{\pi^2}{2\eta^2}, \tag{6}$$

where $\eta = \min(s(a_1,a_2),\ s(f_n(z_0),\ f_n(z_1))$.

On the other hand, we have by Lemma 3,

$$M(f_n(A_\delta)) \geq \int_\delta^\rho \frac{dr}{r\psi^*_{\mu_n}(z_0,r)}. \tag{7}$$

Since

$$\frac{|1-e^{-2i\theta}\mu_n(z_0+ re^{i\theta})|^2}{1-|\mu_n(z_0+ re^{i\theta})|^2} \leq \max\left(1,\ \frac{|1-e^{-2i\theta}\mu(z_0+ re^{i\theta})|^2}{1-|\mu(z_0+ re^{i\theta})|^2}\right),$$

it follows that

$$\psi^*_{\mu_n}(z_0,r) \leq 1 + \psi^*_\mu(z_0,r).$$

Hence, by (7),

$$M(f_n(A_\delta)) \geq \int_\delta^\rho \frac{dr}{r(1 + \psi^*_\mu(z_0,r))} = \alpha_\delta. \tag{8}$$

By the hypothesis, the number α_δ, which does not depend on n, tends to ∞ as $\delta \to 0$.

Combining (6) and (8), we obtain for all sufficiently small values of δ,

$$\eta = s(f_n(z_0),\ f_n(z_1)) \leq \frac{\pi}{(2\alpha_\delta)^{1/2}} .$$

This implies the equicontinuity of the family $\{f_n\}$ at the point z_0.

An obvious modification of the above reasoning yields the result for $z = \infty$. In this case use is made of the hypothesis that the integral diverges as $r_2 \to \infty$. This special assumption is needed only to cover the possibility that the point ∞ belongs to E.

By Ascoli's theorem, there exists a subsequence of the functions f_n which is uniformly convergent in the spherical metric in the whole extended plane. Let f be the limit function. If $A = \{z \mid r_1 < |z-z_0| < r_2\}$, it follows from the above that

$$M(f_n(A)) \geq \int_{r_1}^{r_2} \frac{dr}{r(1 + \psi_\mu^*(z_0,r))} .$$

By hypothesis, the integral on the right is positive. From Lemma 4 we thus conclude that f is a homeomorphism of the extended plane.

Finally, let V be a domain, $\overline{V} \subset -E$. Since $\lim \mu_n = \mu$ in V, it follows from theorems on quasiconformal mappings that $f|V$ is μ-quasiconformal ([2], p. 31 and 197).

6. Necessary existence condition. Using the right hand inequality in Lemma 3 we obtain a necessary condition for the existence of μ-homeomorphisms.

T h e o r e m 3. If E is of σ-finite linear measure and a μ-homeomorphism exists, then for every finite z,

$$\int_{r_1}^{r_2} \frac{\chi_\mu^*(z,r)}{r}\, dr$$

tends to ∞ as $r_1 \to 0$ or as $r_2 \to \infty$.

Proof: Let f be a homeomorphism and $A = \{z \mid r_1 < |z-z_0| < r_2\}$. Then clearly $M(f(A)) \to \infty$ if r_2 is kept fixed and $r_1 \to 0$, or if r_1 is fixed and $r_2 \to \infty$. The validity of the condition in the theorem thus follows immediately from the right-hand inequality (4).

In order to compare the conditions in Theorems 2 and 3, let us consider the special case in which $E = \{\infty\}$. Then, by Theorems 2 and 3,

$$\int_1^\infty \frac{dr}{r\psi_\mu^*(r)} = \infty \tag{9}$$

is a sufficient condition for the existence of a μ-homeomorphism, and

$$\int_1^\infty \frac{\chi_\mu^*(r)dr}{r} = \infty \tag{10}$$

is a necessary one.

If

$$\arg \mu(re^{i\theta}) = 2\theta, \tag{11}$$

then $\psi_\mu^*(r) = (1/D)^*(r) \leq 1$. Hence, condition (9) is fulfilled. It follows that under condition (11), μ-homeomorphisms always exist no matter how rapidly $|\mu(z)| \to 1$ as $z \to \infty$.

Assuming $\arg \mu(re^{i\theta}) = 2\theta + \pi$, we have $\psi_\mu^*(r) = D^*(r)$, $\chi_\mu^*(r) = (1/D)^*(r)$. If. furthermore, D does not depend on θ, it follows that

$$\int_1^\infty \frac{dr}{rD(r)} = \infty$$

is a necessary and sufficient condition for the existence of μ-homeomorphisms.

7. <u>A conformal sewing problem</u>. Let us now assume that E is a Jordan curve. The existence of μ-homeomorphisms is then closely related with the solvability of a conformal sewing problem.

Let $\phi:E\to E$ be a sense-preserving homeomorphism and D_1 and D_2 the two domains bounded by E. We show first how to find a locally quasiconformal homeomorphism of D_1 onto itself with boundary values ϕ.

Let h_1 and h_2 be conformal maps of D_1 onto the upper halfplane such that the function $\psi = h_2 \circ \phi \circ h_1^{-1}$ keeps ∞ fixed. On the real axis ψ is then continuous and increases strictly from $-\infty$ to ∞. From this it follows that the Ahlfors-Beurling mapping

$$\beta_\psi(x+iy) = \frac{1}{2}\int_0^1 [\psi(x+ty)+\psi(x-ty)]dt + \frac{i}{2}\int_0^1 [\psi(x+ty)-\psi(x-ty)]dt \tag{12}$$

is a homeomorphism of the extended plane. It maps the upper halfplane onto itself and coincides with ψ on the real axis. Since ψ is continuous, β_ψ is continuously differentiable outside the real axis, and it has a positive Jacobian there. Therefore, the mapping $w_1 = h_2^{-1} \circ \beta_\psi \circ h_1 : D_1 \to D_1$ has the desired properties: It is a locally quasiconformal homeomorphism (in fact, a diffeomorphism) which agrees with the given homeomorphism $\phi:E\to E$ on the boundary.

With the homeomorphism ϕ we associate a function μ_ϕ as follows: In D_1, μ_ϕ coincides with the complex dilatation of a locally quasiconformal homeomorphism w_1 of D_1 onto itself with boundary values ϕ, and elsewhere $\mu_\phi = 0$. The function μ_ϕ is of course not uniquely determined by ϕ, but the above considerations show that such a function can always be defined. We say that the mapping w_1 generates μ_ϕ.

The function ϕ is said to admit conformal sewing if there exist conformal mappings f_1 and f_2 of D_1 and D_2, respectively, onto two complementary Jordan domains such that $f_2 = f_1 \circ \phi$ on E. The following theorems 4 and 5 show the connection between μ-homeomorphisms and conformal sewing.

T h e o r e m 4. A function ϕ admits conformal sewing if and only if a μ_ϕ-homeomorphism exists.

Proof: Let $w_1 : D_1 \to D_1$ be a locally quasiconformal homeomorphism with boundary values ϕ and complex dilatation μ_1. Let w be a μ_ϕ-homeomorphism, where $\mu_\phi = \mu_1$ in D_1. Set $f_1 = w o w_1^{-1}$, $f_2 = w|D_2$. Then f_1 and f_2 are conformal and $f_2 = f_1 o \phi$ on E. Hence, the condition is sufficient.

Suppose, conversely, that f_i is a conformal mapping of D_i, $i = 1,2$, and $f_2 = f_1 o \phi$ on E. Let w_1 be as above, and set $w = f_1 o w_1$ in D_1, $w = f_2$ in D_2. Then w is locally μ_ϕ-quasiconformal in $D_1 \cup D_2$. Furthermore, on E we have $f_1 o w_1 = f_1 o \phi = f_2$, which shows that w is a μ-homeomorphism.

R e m a r k. It follows from the above theorem that if a μ_ϕ-homeomorphism exists for some choice of μ_ϕ, then a μ_ϕ-homeomorphism exists for all choices of μ_ϕ.

If ϕ admits conformal sewing and f_1, f_2 is a pair of solutions, then $g_1 = h o f_1$, $g_2 = h o f_2$ is also a solution where h is an arbitrary Möbius transformation. If there are no other solutions, we say that the conformal sewing is essentially unique. The same term is used for μ-homeomorphisms: For a given μ, μ-homeomorphisms are said to be essentially unique if they are determined up to Möbius transformations.

T h e o r e m 5. Conformal sewing is essentially unique if and only if μ_ϕ-homeomorphisms are essentially unique.

Proof: Let ϕ admit conformal sewing, f_1, f_2 be a pair of solutions, and $w_1 : D_1 \to D_2$ a locally quasiconformal homeomorphism generating μ_ϕ. As was shown above, the function w, $w = f_1 o w_1$ in D_1, $w = f_2$ in D_2, is a μ_ϕ-homeomorphism.

Suppose first that conformal sewing is essentially unique. Let ω be an arbitrary μ_ϕ-homeomorphism and $h = \omega o w^{-1}$. Then $h o f_1 = \omega o w_1^{-1}$, $h o f_2 = \omega$, which shows that $h o f_1$, $h o f_2$ is a solution for the sewing problem. Hence, h is a linear transformation.

Conversely, let us assume that μ_ϕ-homeomorphisms are essentially unique. Let g_1, g_2 be an arbitrary solution for the sewing problem. Set $h = g_i o f_i^{-1}$ in $f_i(D_i)$, $i = 1,2$. Then $how = g_1 o w_1$ in D_1, $how = g_2$ in D_2. From this we see that how is a μ_ϕ-homeomorphism. Hence, h is a linear transformation.

Let us consider the special case that E is the real axis, D_1 and D_2 the upper and lower halfplane, and $\phi: E \to E$ is normalized by the condition $\phi(\infty) = \infty$. From Theorem 4 we conclude:

C o r o l l a r y. A function ϕ admits conformal sewing along the real axis if and only if a μ_ϕ-homeomorphism exists where μ_ϕ is generated by the Ahlfors-Beurling mapping β_ϕ.

In order to derive more explicit sewing criterions we first appeal to Theorems 2 and 3. It follows that if μ_ϕ satisfies the condition in Theorem 2, then ϕ admits conformal sewing, and if ϕ admits conformal sewing, then μ_ϕ fulfills the condition in Theorem 3. We are thus led to the problem of estimating the complex dilatation μ of the mapping β_ϕ.

Set

$$\alpha(x,y) = \phi(x+y)-\phi(x), \qquad \alpha'(x,y) = \phi(x+y) - \int_0^1 \phi(x+yt)\,dt,$$

$$\beta(x,y) = \phi(x)-\phi(x-y), \qquad \beta'(x,y) = -\phi(x-y) + \int_{-1}^0 \phi(x+yt)\,dt.$$

Direct computation yields, in view of the definition (12),

$$\psi_\mu(x,r,\theta) = \frac{\alpha^2+\beta^2}{\alpha\beta'+\alpha'\beta}\sin^2\theta + \frac{\alpha'^2+\beta'^2}{\alpha\beta'+\alpha'\beta}\cos^2\theta - 2\,\frac{\alpha\alpha'-\beta\beta'}{\alpha\beta'+\alpha'\beta}\cos\theta\sin\theta, \tag{13}$$

where the argument value $x + re^{i\theta}$ on the right hand side has been omitted.

We now want to derive upper estimates for the coefficients in (13) at the point (x,y), $y > 0$. Because α'/α and β'/β are < 1,

$$(14) \qquad \frac{\alpha'^2+\beta'^2}{\alpha\beta'+\alpha'\beta} \leq \frac{\alpha}{\beta} + \frac{\beta}{\alpha}\,, \quad \frac{|\alpha\alpha'-\beta\beta'|}{\alpha\beta'+\alpha'\beta} \leq \frac{\alpha}{\beta} + \frac{\beta}{\alpha}\,.$$

Let us assume that for all values of x,

$$\max\left(\frac{\alpha(x,y)}{\beta(x,y)}\,,\ \frac{\beta(x,y)}{\alpha(x,y)}\right) \leq \rho(y).$$

Then, for $0 < \theta \leq \pi/2$,

$$(15) \qquad \left|\frac{(\alpha'^2+\beta'^2)\cos^2\theta - 2(\alpha\alpha'-\beta\beta')\cos\theta\sin\theta}{\alpha\beta'+\alpha'\beta}\right| \leq 5\rho(r\sin\theta)\cos\theta.$$

The first term in (13) is more difficult to estimate. However, laborious computations can be avoided since the estimates of Reed [4] apply to the present situation, with the following modifications. The constant ρ in Reed's paper must be replaced by

$$\rho_0(y) = \max(\rho(\tfrac{1}{4}\,y),\ \rho(\tfrac{1}{2}\,y),\ \rho(\tfrac{3}{4}\,y),\ \rho(y)),$$

and the majorant $\rho/(\rho+1)$ in Reed's inequality (6) by $(\rho_0(y)+\frac{1}{2})/(\rho_0(y)+1)$. The essential conclusion remains valid: At the point (x,y)

$$(16) \qquad \frac{\alpha^2+\beta^2}{\alpha\beta'+\alpha'\beta} \leq C\rho_0(y),$$

where C is an absolute constant.

From (13), (15), and (16) it follows that

$$(17) \qquad \psi_\mu^*(x,r) \leq C\left(\int_0^{\pi/2} \rho_0(r\sin\theta)\sin^2\theta\, d\,\theta + \int_0^{\pi/2} \rho(r\sin\theta)\cos\theta\, d\,\theta\right).$$

With the help of this estimate the following result can be established:

T h e o r e m 6. Let the function ϕ satisfy a condition

$$\frac{1}{\rho(t)} \leq \frac{\phi(x+t)-\phi(x)}{\phi(x)-\phi(x-t)} \leq \rho(t),$$

where ρ is continuous for $t > 0$ and

$$\rho(t) = O(\log \frac{1}{t}) \text{ as } t\to 0, \ \rho(t) = O(\log t) \text{ as } t\to\infty. \tag{18}$$

Then ϕ admits conformal sewing.

Proof: In this case $\rho_0(t) = O(\log \frac{1}{t})$ for small values of t, $\rho_0(t) = O(\log t)$ for large values of t. Hence, by (17),

$$\psi_\mu^*(x,r) = O(\log \frac{1}{r}), \text{ as } r\to 0, \ \psi_\mu^*(x,r) = O(\log r), \text{ as } r\to\infty.$$

From this we see that the condition in Theorem 2 is fulfilled. A μ_ϕ-homeomorphism thus exists, where μ_ϕ is generated by β_ϕ, and the Corollary to Theorem 4 gives the desired result.

In condition (18) we can obviously replace log by $\log.\log\log.\ldots.\log_n$, where $\log_n$ denotes the n times iterated logarithm. On the other hand, a counterexample of Oikawa [3] shows that Theorem 6 is no longer true if we only assume that for some $\varepsilon > 0$, $\rho(t) = O(t^{-\varepsilon})$ as $t\to 0$, $\rho(t) = O(t^{\varepsilon})$ as $t\to\infty$.

R e f e r e n c e s

[1] A. Beurling and L. Ahlfors: The boundary correspondence under quasiconformal mappings. Acta Math. 96, 125-142 (1956)

[2] O. Lehto und K.I. Virtanen: Quasikonforme Abbildungen. Springer-Verlag, Berlin - Heidelberg - New York (1965)

[3] K. Oikawa: Welding of polygons and the type of Riemann surfaces. Kodai Math. Sem. Rep. 13, 37-52 (1961)

[4] T.J. Reed: Quasiconformal mappings with given boundary values. Duke Math. J. 33, 459-464 (1966)

[5] E. Reich and H.R. Walczak: On the behavior of quasiconformal mappings at a point. Trans. Amer. Math. Soc. 117, 338-351 (1965).

From the Classical Theory of Jacobian Varieties

Henrik H. Martens[1)]

The classical theory of jacobian varieties was initiated by Riemann as a tool for the study of closed Riemann surfaces, or algebraic curves. In this talk I should like to present a brief survey of some of its characteristic features, - mostly discovered by Riemann, - and to illustrate its use through a discussion of two old problems from the theory of curves. Both of these problems are still of considerable interest, and admit reformulations within as abstract a version of algebraic geometry as one cares to consider.

Limitations of personal competence dictate that I formulate my discussion in fairly classical terms. This may not altogether be a disadvantage, however, as it will permit a presentation of results with a minimum of technical prerequisites and complications. Like many other abstract theories, modern algebraic geometry rests on an underlying foundation of solid classical achievements which it seeks to clarify and extend. An understanding of these underlying results may serve as a motivation for the study of contemporary work, and as an introduction to its substance and even sometimes its goals.

Let me begin by reminding you of some basic facts about closed Riemann surfaces, and formulating the two problems I had in mind.

1. Closed Riemann Surfaces

A closed Riemann surface is, by definition, a compact, 1-dimensional, connected, complex-analytic variety. It may be visualized as the surface obtained by attaching a finite number of handles

---

1) Address prepared for the XV Scandinavian Congress of Mathematicians in Oslo, August 1968. Research partly sponsored by the National Science Foundation, Grant GP-6011.

to a sphere and imposing a conformal structure on the result. The number of handles is a topological invariant of the surface denoted by g and referred to as the genus of the surface.

The simplest example is the Riemann sphere (of genus 0). On this surface the expression $(az + b)/(cz + d)$, with suitably selected constants, a, b, c and d, defines a meromorphic function with a single, simple pole. The location of the pole may be arbitrarily prescribed. Any closed Riemann surface of genus 0 admits such functions, and any such function defines a biholomorphic homeomorphism of the surface onto the Riemann sphere. Thus all such surfaces are conformally equivalent. On surfaces of higher genera such functions are topologically impossible.

Riemann surfaces of genus 1 are studied in the theory of elliptic functions. Such a surface may be represented as a torus, or as the surface obtained by identifying opposite sides of a parallellogram in the complex plane. Conformally equivalent surfaces are represented by conformally equivalent parallellograms, and it is well-known that the conformal equivalence classes of these are characterized by a complex parameter related to the complex ratio of the sides of the parallellogram. Thus the family of conformal equivalence classes of closed Riemann surfaces of genus 1 form, in a sense, a space of 1 complex dimension. It can be shown that, given any pair of points on such a surface, there is a meromorphic function whose only singularities are simple poles at the given points.

When the genus is ≥ 2, Riemann showed that the family of conformal equivalence classes of closed Riemann surfaces of genus g form, in a sense, a space of $3g - 3$ complex dimensions. There is a subspace of dimension $2g - 1$ of so-called "hyperelliptic" surfaces, characterized by the property of admitting meromorphic functions whose only singularities are two simple poles. Only one of these poles may be arbitrarily prescribed. All closed Riemann surfaces of genus 2 are hyperelliptic, but in higher genera the hyperelliptic surfaces form a nonempty, proper subspace.

Note that a meromorphic function on a closed Riemann surface displays the surface as a ramified covering of the Riemann sphere, whose number of sheets is equal to the number of poles (counting multiplicities) of the function. Thus the hyperelliptic surfaces are those that can be displayed as two-sheeted coverings of the sphere. It is natural to ask how many sheets are required, in general, to display a surface of given genus. Riemann asserted that a closed Riemann surface of genus $g \geq 2$ can always be displayed as a branched covering of the Riemann sphere with a number of sheets not exceeding $\frac{1}{2}(g + 3)$. The first rigorous proof of the assertion appears to have been given by Meis in 1960.

To give a better account of the situation I must now review some material about divisors. A divisor, D, on a closed Riemann surface, X, is an element of the free abelian group generated by the points of X. Thus a divisor may be represented as a formal sum $D = \sum_{Q \in X} m_Q Q$, where m_Q is an integer, called the <u>multiplicity</u> of Q in D, which is equal to zero for all but a finite number of points $Q \in X$. By the <u>degree</u> of a divisor we mean the sum $\deg(D) = \Sigma\, m_Q$. A divisor is said to be positive $(D \geq 0)$ if $m_Q \geq 0$ for all $Q \in X$. This definition gives a lattice-ordering to the set of divisors.

With a meromorphic function, f, on X we can naturally associate a <u>divisor of zero</u>, $(f)_0$, and a <u>polar divisor</u>, $(f)_\infty$, defined in the obvious manner. By the <u>divisor</u> of f we shall mean the difference $(f) = (f)_0 - (f)_\infty$. Divisors of meromorphic functions are called <u>principal divisors</u>, and form a subgroup. The degree of a principal divisor is 0.

Two divisors, D_1 and D_2, are said to be <u>linearly equivalent</u>, denoted $D_1 \sim D_2$, if their difference, $D_1 - D_2$, is principal. This is clearly an equivalence relation. Linearly equivalent divisors have the same degree. If D is a positive divisor, the set of all positive divisors linearly equivalent to D is called the <u>complete linear series</u> determined by D.

If D is any divisor on X, we denote by $L(D)$ the set of all meromorphic functions f on X such that $D + (f) \geqq 0$. To see what this condition means, write $D = D_1 - D_2$, where D_1 and D_2 are positive divisors without common points. Then the inequality $D_1 - D_2 + (f)_0 - (f)_\infty \geqq 0$ can be satisfied only if the contribution from $(f)_\infty$ is cancelled by D_1, and that from D_2 is cancelled by $(f)_0$. We may express this by saying that $f \in L(D)$ if and only if it has zeros at least as "prescribed" by D_2, and poles at most as "allowed" by D_1. In particular, if $D \geqq 0$ then $L(D)$ consists of those functions which have poles at most as "allowed" by D, and will always contain all constant functions. If $\deg(D) < 0$, $L(D)$ is empty since we then have "prescribed" more zeros than we have "allowed" poles.

Since the addition of two meromorphic functions preserves common zeros and does not introduce new poles, $L(D)$ has a natural structure as a linear space. The dimension of $L(D)$ is denoted by $\ell(D)$. If $D \geqq 0$, then $\ell(D) \geqq 1$ and we denote by $\dim(D)$ the integer $\ell(D) - 1$. We refer to $\dim(D)$ as the <u>dimension</u> of D, and note that the dimension of a positive divisor is strictly positive if and only if there are non-constant meromorphic functions in $L(D)$. Linearly equivalent divisors have the same dimension, and by the <u>dimension</u> and <u>degree</u> of a complete linear series we shall mean the dimension and degree of any of its divisors.

If D is a positive divisor of degree n and dimension $\geqq r$, let $f_0, f_1, \ldots, f_r$ be a linearly independent set of functions in $L(D)$. By solving an obvious set of linear equations, we can determine a non-vanishing linear combination of the f_i whose divisor of zeros contains an arbitrarily prescribed r-tuple of points (other than those in D). Hence, if $\dim(D) \geq r$, then the complete linear series determined by D contains positive divisors with arbitrarily prescribed r-tuples of points. The converse is easily established also. From this we immediately get $\dim(D) \leq \deg(D)$, and hence $L(D)$ is finite-dimensional. (If D_1 and D_2 are positive divisors, then $L(D_1 - D_2) \subset L(D_1)$, hence $L(D)$ is finite-dimensional for arbitrary D).

By a <u>holomorphic differential</u> on a closed Riemann surface X we mean a closed 1-form which locally is the differential of a holomorphic function. By the <u>divisor</u> of a holomorphic differential we mean the divisor determined in the obvious way by its zeros, and such divisors are called <u>canonical divisors</u>. If Z is a canonical divisor, we have $\ell(Z) = g$ and $\deg(Z) = 2g - 2$. The quotient of two holomorphic differentials is a meromorphic function, and hence all canonical divisors form a complete linear series known as the <u>canonical series</u>. If D is a positive divisor we denote by $i(D)$ the dimension of the linear space of holomorphic differentials whose divisors contain D.

As a reward for your patience in following all these definitions I now give you the Riemann-Roch theorem for positive divisors:

$$\dim(D) = \deg(D) + i(D) - g,$$

which contains a wealth of information. Note first that since $i(D) \geqq 0$, we always have $\dim(D) \geqq \deg(D) - g$. From this we can immediately verify our earlier remarks about the existence of meromorphic functions with exactly 1 and 2 poles on surfaces of genus 0 and 1, respectively. If the genus is at least 2, then $2g - 2 \geq g$, and holomorphic differentials have at least g zeros. By a proper choice of D of degree g we may then insure that $i(D) \geq 1$, and hence $\dim(D) = i(D) \geq 1$. Thus we establish the existence of nonconstant meromorphic functions with at most g poles. In particular, there are functions with precisely 2 poles in genus 2.

Since holomorphic differentials have $2g - 2$ zeros, we get $i(D) = 0$ if $\deg(D) > 2g - 2$. For positive divisors of degree $\leqq 2g - 2$ it may happen, however, that $i(D) \geqq 1$. Such divisors are called _special divisors_. For special divisors we have a consequence of the Riemann-Roch theorem known as Clifford's theorem:
_If_ D _is a special divisor, then_ $2 \dim(D) \leqq \deg(D)$, _and if_ $0 < \deg(D) < 2g - 2$ _then the inequality is always strict except on a hyperelliptic surface._[1]

By the Riemann-Roch theorem we have $\deg(D) - \dim(D) = g - i(D) \leqq g - 1$ when D is special, and combining this with Clifford's theorem we conclude that for special divisors

$$0 \leqq \dim(D) \leqq \deg(D) - \mathrm{Dim}(D) \leqq g - 1.$$

Consider now the complete linear series determined by a special divisor. If its dimension is r and its degree is $r + s$, then we have $0 \leqq r \leqq s \leqq g - 1$. Riemann asserted that the set of complete linear series of dimension r and degree $r + s$ forms, in a sense, a space of $(r + 1)s - rg$ complex dimensions. In particular, if $r = 1$, we find that the set of complete linear series of dimension 1 and degree $s + 1$ forms a space of dimension $2s - g$. Hence if $2s - g \geqq 0$, there are non-constant meromorphic functions with at most $s + 1$ poles. This is the result of Meis referred to earlier. Riemann's general assertion has, to my knowledge, never been rigorously established in spite of many attempts. The proof of Meis is restricted to the classical case of Riemann surfaces, but the assertions are perfectly meaningful (and plausible) in the theory of algebraic curves over fields of arbitrary characteristic, and there the question is open. This is the first of the two problems I wanted to discuss.

---

[1] Phil. Trans. 169(1878), 681. For a simple proof see Crelle's Journal, Vol. 233 (1968) p. 92.

Let me now turn to a different matter. It is known that the first homology group of a surface of genus g is a free abelian group on $2g$ generators. It is also known that one can, in several ways, select a so-called canonical basis, $\alpha_1,\alpha_2,\dots,\alpha_{2g}$ characterized by the intersection matrix

$$(\alpha_i \circ \alpha_j) = \begin{pmatrix} 0 & E \\ -E & 0 \end{pmatrix}$$

where E denotes the unit $g \times g$ - matrix. (The elements of a canonical basis correspond to the canonical retrosections of Riemann surfaces represented as coverings of the sphere).

Let $\omega^1,\dots,\omega^g$ be a basis for the holomorphic differentials on a closed Riemann surface of genus g. Let $\alpha_1,\dots,\alpha_{2g}$ be a canonical homology basis, and consider the $g \times 2g$ matrix with entries $\int_{\alpha_j} \omega^i$. Riemann showed that with an appropriate choice of the ω^i, the matrix becomes $(\pi i E, A)$, where A is a symmetric $g \times g$ matrix whose real part is negative definite. A different choice of canonical basis will generally result in a different matrix A, but the matrices obtained from a given Riemann surface are related by transformations connected with Siegel's symplectic modular group. The group is discontinuous, and the different matrices form a discrete set.

Notice now that the set of matrices A satisfying the given conditions form a space of dimension $\frac{1}{2}g(g+1)$. If $g \geqq 4$, this is strictly greater than the number $3g - 3$ given by Riemann as the dimension of the space of conformal equivalence classes of closed Riemann surfaces of genus g. One naturally asks whether a characterization can be given for those matrices which occur as period matrices. This is the second problem I wanted to discuss, and as I have just given one characterization, you will note that it is rather vaguely formulated. If one has in mind something like a set of equations in the matrix entries, the history of the problem can be written in two sentences: The first contribution was a notoriously obscure paper by Schottky in 1888, giving a necessary condition in genus 4. The second contribution was a paper by Andreotti and Mayer in 1967, giving necessary conditions for $g \geqq 4$.

But at this point it is impossible to continue the discussion without first introducing the jacobian variety of a closed Riemann surface.

2. Jacobian Varieties

Let X be a closed Riemann surface of genus $g \geq 1$. Let $\omega^1,\dots,\omega^g$ be any basis for the holomorphic differentials on X, and let $\alpha_1,\dots,\alpha_{2g}$ be a homology basis. If Π is the period matrix formed with these bases, its column vectors, $\pi_1,\dots,\pi_{2g}$, are $\mathbb{R}$ - linearly independent vectors in $\mathbb{C}^g$. We denote by $\mathbb{C}^g/\Pi$ the quotient space obtained by identifying points of $\mathbb{C}^g$ which differ by an integral linear combination of the π_i. Clearly $\mathbb{C}^g/\Pi$ has a natural structure as an additive group and as a compact, complex-analytic variety.

Select a point $P \in X$ as a reference, and for any $Q \in X$ select a path joining P and Q. Let $\hat{u}(Q)$ be the point in $\mathbb{C}^g$ defined by $\hat{u}^i(Q) = \int_P^Q \omega^i$, where the integral is taken along the selected path. Different choices of the path will determine points $\hat{u}(Q)$ which differ by an integral linear combination of the π_i, and hence the procedure determines a unique point $u(Q) \in \mathbb{C}^g/\Pi$. Thus we have defined a map $\varkappa : X \to \mathbb{C}^g/\Pi$ which is clearly holomorphic.

The map $\varkappa$ can be extended by setting $\varkappa(D) = \Sigma\, m_Q \varkappa(Q)$ for any divisor $D = \Sigma\, m_Q Q$ in X. With this notation we get the following version of Abel's theorem:

Two positive divisors D_1 and D_2 on X of the same degree are linearly equivalent if and only if $\varkappa(D_1) = \varkappa(D_2)$.

As a result we see that the positive divisors of a complete linear series on X are all mapped on the same point, and *the set of complete linear series of a given degree on X are in a 1 - 1 correspondence with a subset of $\mathbb{C}^g/\Pi$.*

Denote by $G^r{}_n$ the set of points of $\mathbb{C}^g/\Pi$ which are images of positive divisors of degree n and dimension $\geqq r$. The assertion of Riemann referred to earlier can now be formulated as an assertion about the dimension of the set $G^r{}_n$, and the following result can be obtained:

Let $0 \leqq r \leqq s \leqq g - 1$. Then $G^r{}_{r+s}$ is an analytic subset of $\mathbb{C}^g/\Pi$ and every non-empty component of $G^r{}_{r+s}$ has dimension at least $(r + 1)s - rg$. Note, however, that we do not assert that $G^r{}_{r+s}$ has any non-empty component.

Let me indicate how the result can be obtained. First some notation: We denote by W^s the set of images of positive divisors of degree $\leqq s$. If S is a subset of $\mathbb{C}^g/\Pi$, and if v is a point in $\mathbb{C}^g/\Pi$, we denote by S_v the set of points $s + v$, with $s \in S$, and refer to S_v as the *translate of* S *by* v.

We have already shown that if D is a positive divisor of degree n and dimension r, then for every positive divisor D' of degree r there is a positive divisor D'' of degree $n - r$ such that $D \sim D' + D''$, and conversely. Translated into the present notation we find that if $u \in G^r{}_n$, then for every $v \in W^r$ there is a $w \in W^{n-r}$ such that $u = v + w$. Equivalently, $u \in W^{n-r}_{v}$ for every $v \in W^r$. Thus we get

$$G^r{}_n = \cap \{ W^{n-r}_{v}, v \in W^r \}.$$

Now, W^s is the image in $\mathbb{C}^g/\Pi$ of the s-fold Cartesian product X^s under $\varkappa$, and it isn't hard to believe that this is an analytic subset of $\mathbb{C}^g/\Pi$. But then $G^r{}_n$, being an intersection of analytic subsets, is itself analytic. It is easily shown that the dimension of W^s is s, and by appropriately choosing the points $v \in W^r$, one can show that G^r_n is a component of an intersection of a certain finite number of sets W^{n-r}_v. Its dimension can then be estimated by the well-known formulas. The trouble is that the finite intersection contains other components, and hence we cannot conclude that G^r_n exists, even when the formula yields a non-negative dimension. On the other hand, one can get additional information. In the present notation, Clifford's theorem says that if $0 < r < g - 1$, then G^r_{2r} is non-empty, i.e. $\dim(G^r_{2r}) = 0$, only if X is hyperelliptic. The following generalization can be proved:

Let $0 \leqq r \leqq s \leqq g - 1$. _Then_ $\dim(G^r_{r+s}) \leqq s - n$, _and if_ $0 < r \leqq s < g - 1$ _then_ $\dim(G^r_{r+s}) = s - r$ _if and only if_ X _is hyperelliptic_.

Finally, the dimension formula of Riemann can be sharpened slightly as follows:

Let $0 \leq r \leq s < g - 1$. _Then_, if $G^r_{r+s} \neq \emptyset$

$$\dim_*(G^r_{r+s}) \geqq \dim_*(G^r_{r+s+1}) - (r + 1)$$

where $\dim_*$ denotes the dimension of the smallest non-empty component. Note that we have $G^r_{r+s} \subset G^r_{r+s+1}$, and hence it suffices to assume $G^r_{r+s} \neq \emptyset$.

The preceding should suffice to establish the importance and utility of $\mathbb{C}^g/\Pi$. Let us now remove the accidents of the construction. Consider the group of divisors on X of degree 0. The map $\varkappa$ gives a homomorphism of this group into $\mathbb{C}^g/\Pi$, and we observe that this homomorphism is independent of the choice of reference point $P \in X$. Abel's theorem asserts that the kernel of this homomorphism is the subgroup of principal divisors. Finally, the solution of the classical Jacobi inversion theorem asserts that the homomorphism is onto. Thus $\mathbb{C}^g/\Pi$ is canonically identified with a quotient of the group of divisors of degree 0, endowed with a complex-analytic structure. Thus the choices involved in its construction are non-essential, and we have defined the _jacobian variety_, $J(X)$, of X.

(The map $\varkappa = X \to J(X)$ can be defined by the formula $Q \to (Q - P)$). From this point of view the jacobian variety is a special case of the Picard variety of a complex-analytic variety.

There is a second way of looking at $J(X)$ which is of some interest. The quotient $\mathbb{C}^g/\Pi$ is an example of a complex torus, by which we mean a quotient space $\mathbb{C}^n/G$, where G is a discrete group of translations generated by $2n$ $\mathbb{R}$-linearly independent vectors in $\mathbb{C}^n$. The matrix formed with these vectors as column vectors will be called a period matrix for $\mathbb{C}^n/G$, and the integral linear combinations of the vectors will be referred to as periods.

Denoting the coordinates of $\mathbb{C}^n$ by $\hat{u}^1,\ldots,\hat{u}^n$, it is not hard to show that their differentials induce a basis, $du^1,\ldots,du^n$ for the holomorphic differentials on $\mathbb{C}^n/G$. Now, suppose V is a complex-analytic variety, and $\varphi : V \to \mathbb{C}^n/G$ is a holomorphic map. Let $\omega^1,\ldots,\omega^n$ be the holomorphic differentials on V obtained by pulling back the du^i via φ. One easily sees that φ may be obtained by first mapping V into $\mathbb{C}^n$ by a formula

$$\hat{u}^i(Q) = \hat{u}^i(P) + \int_P^Q \omega^i$$

and then projecting the result on $\mathbb{C}^n/G$. In particular, if V is itself a complex torus $\mathbb{C}^m/F$, then any holomorphic map $\varphi : \mathbb{C}^m/F \to \mathbb{C}^n/G$ must be of the form $u = L(v) + u_0$, where u_0 is a constant in $\mathbb{C}^n/G$, and L is a group homomorphism. Note that L can be represented by an $n \times m$ matrix which takes periods of $\mathbb{C}^m$ into periods of $\mathbb{C}^n$.

Consider now a holomorphic map $\varphi : X \to \mathbb{C}^n/G$ such that $\varphi(P) = 0$. It is clear from the preceding that φ can be represented as a composite map

$$X \xrightarrow{\varkappa} J(X) \xrightarrow{\varphi^*} \mathbb{C}^n/G,$$

where φ^* is a uniquely determined homomorphism of complex tori. This defines $J(X)$ up to a canonical isomorphism, and from this point of view the jacobian variety is a special case of the Albanese variety of a complex-analytic variety.

The jacobian varieties of closed Riemann surfaces are in a special class of complex tori known as abelian varieties, characterized by the property that they can be embedded as algebraic subvarieties in a suitable projective space. It can be shown that a condition for a complex torus to be an abelian variety is that its period matrix satisfies conditions similar to Riemann's relations on a period matrix of a Riemann surface. When these conditions fail to be met, one can get complex tori which admit no non-trivial meromorphic functions at all. This leads us naturally to the topic of theta-functions or multiplicative functions.

3. Multiplicative Functions ([7],[11],[20])

Let us assume that the jacobian variety of X is represented by $J(X) = \mathbb{C}^g/\Pi$ where $\Pi = (\pi i E; A)$ is a canonical period matrix of X. Since A has a negative definite real part, it can be shown that the series

$$\theta(\hat{u};A) = \sum_{m \in \mathbb{Z}^g} e^{{}^t m(Am + 2\hat{u})}$$

converges absolutely and uniformly on compact subsets of $\mathbb{C}^g$, and thus defines a holomorphic function. A little calculation shows that

$$\theta(\hat{u} + \pi i e_h; A) = \theta(\hat{u};A)$$

and

$$\theta(\hat{u} + a_h; A) = \theta(\hat{u};A)e^{-(2\hat{u}^h + a^h{}_h)}$$

where e_h is the h-th column vector of E, and a_h is the h-th column vector of A. Since the exponential factor in the second equation never vanishes, the zeros of $\theta(\hat{u};A)$ depend only on the projection of $\hat{u}$ in $J(X)$.

$\theta(\hat{u};A)$ is known as a first-order theta-function over $J(X)$. Other first-order theta-functions can be obtained by multiplying $\theta(\hat{u} - \hat{v}_0;A)$ by a normalizing exponential factor, $\hat{v}_0$ being a constant. Higher-order theta-functions can be obtained by taking products of first order functions, and by taking suitable quotients of higher-order functions one obtains periodic functions which define meromorphic functions on $J(X)$. Theta- functions will also yield embeddings of $J(X)$ in projective space. But we shall here confine our attention to the first order functions.

We have already observed that $\theta(\hat{u};A)$ induces a well-defined set of zeros on $J(X)$. Riemann made the amazing discovery that <u>this set of zeros is precisely a translate of</u> W^{g-1}, [20] The proof is essentially a straight-forward, but somewhat tedious, exercise in "herumintegrieren", i.e. integration around a polygon representing the Riemann surface X.

Note first that if the surface is cut up to form a polygon, the integral formula for the map $\varkappa$ yields a map of the polygon into $\mathbb{C}^g$, and a branch of $\theta(\hat{u};A)$ may be pulled back on the polygon. Its zeros may be counted by integrating the logarithmic derivative around the polygon. Taking into account the periodicity behavior of θ, and assuming that it doesn't vanish identically on the polygon, one computes the number of zeros as g.

Next, the coordinate functions $\hat{u}^i$ may be pulled back on the polygon and by a similar procedure we can evaluate the sum $\sum_j \hat{u}^i(Q_j)$ where Q_j are the zeros induced by θ on X. The results can be summarized as follows:

<u>Let</u> u_1, u_2 <u>be points of</u> $J(X)$ <u>such that</u> $\theta(\hat{u} - \hat{u}_1;A)$ <u>and</u> $\theta(\hat{u} - \hat{u}_2;A)$ <u>do not vanish identically over</u> $\varkappa(X) = W^1 \subset J(X)$. Let D_1 and D_2 <u>be the divisors induced on</u> X <u>by these functions</u>. <u>Then</u>

$$\deg(D_1) = \deg(D_2) = g$$

and

$$\varkappa(D_1) - \varkappa(D_2) = u_1 - u_2$$

Suppose now that we keep u_2 fixed, and let u_1 vary over $J(X)$. Then the left hand side varies over a translate of W^g, and one easily concludes that $W^g = J(X)$. This was Riemann's solution of the Jacobi inversion problem.

Next, if u_1 is such that $\theta(-\hat{u}_1;A) = 0$ then $D_{u_1} = D'_{u_1} + P$, where $\deg(D'_{u_1}) = g - 1$. When u_1 varies over such values, the left hand side varies over a translate of W^{g-1}. Of course one has to attend to some details arising from the possibility that $\theta(\hat{u} - \hat{u}_1;A)$ vanishes identically over W^1, but the above outline gives the main idea of the proof.

I shall now give a more general formulation. Let $\mathbb{C}^n/G$ be a complex torus, and let Ω be a period matrix for $\mathbb{C}^n/G$ with column vectors $\omega_1,\ldots,\omega_{2n}$. Let Λ be an $n \times 2n$ matrix with column vectors $\lambda_1,\ldots,\lambda_{2n}$, and let $\gamma = (\gamma^1,\ldots,\gamma^{2n})$ be a 2n-tuple of complex constants. A holomorphic function Ψ on $\mathbb{C}^n$ is said to be <u>multiplicative of type</u> (Ω,Λ,γ) <u>over a subset</u> $S \subset \mathbb{C}^n/G$ if, for every $\hat{u} \in \mathbb{C}^n$ whose projection lies in S,

$$\Psi(\hat{u} + \omega_j) = \Psi(\hat{u}).\ e^{2\pi i({}^t\lambda_j\hat{u} + \gamma^j)}$$

As in the case of the theta-function, Ψ doesn't define a function on $\mathbb{C}^n/G$ or S, but it induces a well-defined set of zeros on S. The theta-function is an example of a function multiplicative over the whole torus, but if we consider any first-order partial derivative $\partial\theta$ of θ, we get a function which is multiplicative only on a proper subset (the zeros of θ) of the torus.

In order for the above formula to be possible, Ω and Λ must satisfy a compatibility condition. Applying the formula to $\Psi(\hat{u} + \omega_j + \omega_k)$, we get

$\Psi(\hat{u} + \omega_j + \omega_k) = \Psi(\hat{u})e^{2\pi i({}^t\lambda_j u + \gamma^j)}p^{2\pi i({}^t\lambda_k(\hat{u} + \omega_j) + \gamma^k)}$

and

$$\Psi(\hat{u} + \omega_k + \omega_j) = \Psi(\hat{u})e^{2\pi i({}^t\lambda_k\hat{u} + \gamma^k)}e^{2\pi i({}^t\lambda_j(\hat{u} + \omega_k) + \gamma^j)}$$

whence $e^{2\pi i({}^t\lambda_j\omega_k - {}^t\lambda_k\omega_j)} = 1$. It follows that the skew-symmetric matrix

$$N = {}^t\Omega\Lambda - {}^t\Lambda\Omega$$

must have integral entries. N is known as the <u>characteristic matrix of Ψ with respect to Ω</u>.

Suppose now that $\varphi : X \to \mathbb{C}^n/G$ is a holomorphic map. If Ψ is multiplicative of some type over the image of X in $\mathbb{C}^n/G$, it either vanishes identically over $\varphi(X)$, or else induces a well-defined set of zeros corresponding to a positive divisor on X.

We have already shown that any such map $\varphi : X \to \mathbb{C}^n/G$ can be factored through $J(X)$:

$$X \xrightarrow{\kappa} J(X) \xrightarrow{\varphi^*} \mathbb{C}^n/G.$$

Hence Ψ may be pulled back as a multiplicative function over W^1 in $J(X)$. To study the divisor induced by Ψ on X, it therefore suffices to examine the case when $\mathbb{C}^n/G = J(X)$ and $\varphi = \kappa$.

Now, let Π be a period matrix for X formed with a canonical homology basis, and let Ψ be a holomorphic function on $\mathbb{C}^g$ which is multiplicative of type (Π,Λ,γ) over a subset $S \subset J(X)$. Then the function Ψ_v defined by $\Psi_v(\hat{u}) = \Psi(\hat{u} - \hat{v})$ will be multiplicative of type $(\Pi,\Lambda,\gamma - {}^t\Lambda\hat{v})$ over S_v. Consider now those $v \in J(X)$ such that $W^1 \subset S_v$. For each such v, Ψ_v is multiplicative over W^1, and if it doesn't vanish identically over W^1, it induces a well-defined divisor D_v on X. It turns out that $\deg(D_v)$ is independent of v, and that

$$\kappa(D_v) - \kappa(D_o) = L(v)$$

where L is an endomorphism of $J(X)$.

To get explicit formulas, let me make a remark about the endomorphisms of $J(X)$. Consider the map $\mathbb{R}^{2g} \to \mathbb{C}^g$ given by $\hat{u} = \Pi\hat{x}$, where $\hat{u} \in \mathbb{C}^g$ and $\hat{x} \in \mathbb{R}^{2g}$ are column vectors. This induces an isomorphism

$$\mathbb{R}^{2g}/\mathbb{Z}^{2g} \to \mathbb{C}^g/\Pi = J(X)$$

where $\mathbb{R}^{2g}/\mathbb{Z}^{2g}$ is a real 2g-dimensional torus. Any endomorphism of $\mathbb{C}^g/\Pi$ is given by a $g \times g$ matrix T acting on $\mathbb{C}^g$ such that periods go into periods. But then $T\Pi = \Pi M$, where M is a $2g \times 2g$ matrix with integral entries. Moreover, M defines an endomorphism of the underlying real torus $\mathbb{R}^{2g}/\mathbb{Z}^{2g}$ which uniquely represents the given endomorphism of $\mathbb{C}^g/\Pi$. With this we can now give a precise formulation of the results:

Let Π be a period matrix of X formed with a canonical homology basis. Let Ψ be a holomorphic function on $\mathbb{C}^g$ which is multiplicative of type (Π,Λ,γ) over a subset $S \subset J(X)$. Assume that $W^1 \subset S$, and that Ψ doesn't vanish identically over W^1. For any $v \in J(X)$ such that $W^1 \subset S_v$ and Ψ_v doesn't vanish identically over W^1, let D_v be the divisor induced on X by Ψ_v.

Then
$$\varkappa(D_v) - \varkappa(D_0) = L(v)$$

where L is an endomorphism of $J(X)$ represented on the underlying real torus by the matrix

$$M = JN$$

and
$$\deg(D_v) = \tfrac{1}{2}\,\mathrm{Trace}\,(JN).$$

Here $N = {}^t\Pi\Lambda - {}^t\Lambda\Pi$ is the characteristic matrix of Ψ, and $J = \begin{pmatrix} 0 & E \\ -E & 0 \end{pmatrix}$.

4. Applications

Let us now take a second look at the problems posed in the introduction. We have already shown that the sets $G^r_n \subset J(X)$ form a natural setting for the study of the set of complete linear series on X. Riemann's characterization of W^{g-1} as the set of zeros of a first order theta-function, shows that W^{g-1} is an analytic subset of $J(X)$. It is not hard to see that the sets W^s, and hence also the sets $G^r{}_n$, can be obtained by intersecting suitable translates of W^{g-1}, and thus the analytic character of all these sets is established. In fact, if $J(X)$ is represented as an algebraic sub-variety of some projective space, all these sets turn out to be algebraic subsets.

Riemann's studies of the vanishing of the theta-functions lead to another characterization of the sets $G^r{}_n$. If we denote by θ_K the first-order theta-function whose zeros are W^{g-1}, Riemann showed that θ_K <u>has a zero of multiplicity</u> $m \geqq 1$ <u>over</u> $u \in J(X)$ <u>if and only if</u> $u = \varkappa(D)$ <u>where</u> $\deg(D) = g - 1$ <u>and</u> $\ell(D) = m$, D <u>being a positive divisor on</u> X.

A vague idea of the proof may be obtained by observing that $W^1 \subset W^{g-1}$, and hence θ_K vanishes identically over W^1. But then any first-order partial derivative of θ_K is multiplicative over W^1 of the same type as θ_K (differentiate the periodicity formulas). It follows that the quotient of two such derivatives is periodic over W^1, and hence defines a meromorphic function on X. The divisors induced by such derivatives on X are therefore linearly equivalent. Clearly, the induced divisors have degree g, since the multiplicative type is the same. But if 0 is a multiple zero of θ_K, all first-order derivatives vanish at 0, and the induced divisors have a common point. Thus we effectively have linear equivalence between divisors of degree $g - 1$.

The early proofs of Riemann's result were often lacunary. For recent expositions one may consult [11] and [15]. A proof of the result along the above lines was given in [12]. As a byproduct one can show that if D is any special divisor on X, then $L(D)$ is generated by the quotients of suitable selected first-order partial derivatives of theta-functions. This remark is related to earlier results of Christoffel [6]. I understand that Mumford has an unpublished proof of the result valid in arbitrary characteristic.

Note that the preceding enables us to identify G^1_{g-1} with the set of singular points of W^{g-1}. Weil [24] showed that, in general, G^1_n is the set of singular points of W^n, $n \leqq g-1$, and this result is true in the abstract case. In his abstract theory of abelian varieties, Weil succeeded in reformulating and often improving most of the key results of the classical theory. Thus, the results about endomorphisms defined by multiplicative functions is extended to intersections of subvarieties of complementary dimensions. For expositions see [24] and also [10].

I mentioned earlier that Meis [16] proved (classical case) that $G^1_{s+1} \neq \emptyset$ when $2s \geqq g$. Thus W^{s+1} will always have singular points when $2s \geqq g$. Now W^{s+1} is the holomorphic image of the $(s+1)$ - fold symmetric product $X^{((s+1))}$, obtained as a quotient space of the cartesian product $X^{(s+1)}$ by the symmetric group. One can show (e.g. [2]) that $X^{((s+1))}$ has a (non-singular) complex-analytic structure, and the map fails to be $1-1$ exactly at points corresponding to positive divisors of positive dimension. Meis' result is therefore equivalent to the statement that the $(s+1)$ - _fold symmetric product of a closed Riemann surface of genus_ $g \geqq 1$ _cannot be holomorphically imbedded in a complex torus when_ $2s \geqq g$. As mentioned before, the corresponding result in the abstract case is an open problem (so far as I know). An idea of Riemann's leads to a very simple proof that $G^1_{s+1} \neq \emptyset$ when $s \geqq 3/4\, g - 1$, and this proof is valid for fields of characteristic $\neq 2$, [13].

Let us now turn to the problem of period relations. Observe first that if $\mathbb{C}^n/G$ is an arbitrary complex torus whose period matrix may be taken as $(\pi i E; A)$ where A satisfies the Riemann relations, then a theta-function may be defined. We shall refer to such tori as <u>abelian varieties</u>. These can all be imbedded as algebraic subvarieties of some projective space. The problem of characterizing those Riemann matrices which come from Riemann surfaces, let us call them <u>special</u> Riemann matrices, can therefore be replaced by that of characterizing <u>special</u> theta-functions, or that of characterizing jacobian varieties among abelian varieties. The last problem is the meaningful formulation in the abstract case.

Let me first point out a pitfall. It is essential that the matrix $(\pi iE;A)$ should be a canonical matrix, i.e. formed with a canonical homology basis, of X. If we change to a different canonical homology basis, we get a matrix of the same form, and the two are related by an element of Siegel's symplectic modular group. It is perfectly possible, however, that two matrices $(\pi iE;A_1)$ and $(\pi iE;A_2)$ define isomorphic tori, but are not related by a symplectic transformation. Thus $(\pi iE;A_1)$ may be a canonical matrix for X, while $(\pi iE;A_2)$ is not. Then the theta-functions formed with A_2 will not satisfy the theory outlined before. For instance, if A_2 is a direct sum of submatrices

$$A_2 = \begin{pmatrix} A'_2 & 0 \\ 0 & A_2'' \end{pmatrix}$$

one easily sees that the theta-function formed with A_2 splits into a product of two theta-functions in lower-dimensional spaces. The set of zeros induced by such a theta-function on the torus is then a reducible set, and could not be the set W^{g-1} of a jacobian variety, which is irreductible. Now, Weil has an (unpublished) example of a jacobian variety of dimension 4 which has a period matrix $(\pi iE;A)$ where A splits. This cannot then be a canonical period matrix for the Riemann surface, despite the fact that A satisfies the Riemann period relations. Hayashida and Nishi [8] have shown that there is a large class of such examples in dimension 2. It is also possible that a given abelian variety has period matrices $(\pi iE;A_1)$ and $(\pi iE;A_2)$ which are canonical period matrices of distinct Riemann surfaces, - such an example is quoted by Severi [22].[1)]

On the other hand, a theorem of Torelli [23] shows that if $(\pi iE;A_1)$ and $(\pi iE;A_2)$ are symplectically related, and if both are canonical period matrices of Riemann surfaces, then the Riemann surfaces are conformally equivalent.

From this we see that the proper objects to study are not just abelian varieties, but what Weil calls _polarized abelian varieties_, i.e. abelian varieties with distinguished period matrices. The notion of polarization carries into the abstract theory, replacing theta-functions by the equivalent of their set of zeros. (Mumford [17] has recently obtained an equivalent of theta-functions in the abstract theory also.) An intrinsic characterization of polarized jacobian varieties was given by Matsusaka [14]. With this characterization, Baily [4] showed that the special Riemann matrices lie on an irreducible analytic subset of dimension $3g - 3$ of the space of Riemann matrices from which a lower-dimensional analytic subset has been deleted. It appears to follow from results of Hoyt [9] that the set to be deleted is precisely that which corresponds to matrices which split into a direct sum, and their symplectic equivalence classes. Weil [25] had earlier shown that any 2×2 Riemann matrix is special, except (those symplectically equivalent to) the diagonal ones. A similar result should hold in dimension 3, by Hoyt's results.

The idea of characterizing special Riemann matrices by properties of the associated theta-functions, is an old one. Schottky [21] derived a polynomial in the values taken by certain first-order

---

1) One could even conceive of the possibility that A is carried into itself by a non-symplectic transformation. Then the torus would have two essentially distinct representations as the Jacobian variety of the same surface. I know of no example of this, however.

theta-functions of 4 variables at the origin, and showed that this polynomial vanishes if the Riemann matrix is special, but is not identically zero as a function of Riemann matrices. Thus he defined an analytic subset of the space of Riemann matrices, one component of which must be the special matrices (apart from a lower-dimensional subset). Andreotti and Mayer [3] obtained relations for all dimensions ≥ 4. In contrast to the obscurity of Schottky's paper, the Andreotti-Mayer approach is based on a beautifully simple idea. We have already shown that G^1_{g-1} is the set of singularities of W^{g-1}. Hence for $g \geq 4$, the special theta-functions must be singular over a set of dimension $\geq g - 4$. Andreotti and Mayer succeeded in showing that the presence of such singularities leads to explicit analytic conditions on the Riemann matrix which define an analytic sub-space of dimension $3g - 3$. Again the special matrices must constitute a component of this set.

Notice, however, that in both papers only necessary conditions have been established. Whether the Schottky condition is sufficient (modulo the subsets to be deleted) is not known. It would be, if the analytic set defined is irreducible. It is easy to see that the Andreotti-Mayer set is reducible in genus 4 (and I doubt that it is irreducible in higher genera). For those who understand the language, it suffices to remark that the condition that an even theta-constant vanishes defines at least onecomponent of the space of Riemann matrices whose theta-functions have singularities. The classical classification of surfaces of genus 4 shows that the special matrices intersect such a component in a lower-dimensional subset. Thus there are at least some $\frac{1}{2}(2^8 + 2^4) = 136$ unwanted components.

I should mention also that Rauch and Farkas [19] have found an "ansatz" which promises to lead to explicit period relations. Rauch has done important work on the problem of moduli in the context of Teichmüller spaces (see [18]). This carries me outside the framework of my talk, but I should at least mention the beautiful contributions of Ahlfors and Bers to this theory (see [1] and [5]). One may note that Meis' work referred to earlier, relies very heavily on this theory.

Let me conclude by mentioning a third problem. In genus 2, the zeros of the theta-function are precisely $W^{2-1} = W^1$, i.e. the imbedded image of X. While it can be shown that W^1 is an intersection of translates of W^{g-1} for $g > 2$, I know of no way of picking the translates without referring to W^1. The problem of reconstructing W^1 from W^{g-1}, (and hence from the theta-function, (and hence from the matrix)), seems to me to be the <u>true</u> inversion problem of the theory, however vaguely I may have stated it. At any rate, I hope to have shown that there is plenty of work left in this fascinating area.

Bibliography (This list contains only works to which explicit reference is made in the text.)

1. L. V. Ahlfors, Teichmüller Spaces, Proc. Int. Congr. Math., Stockholm 1962, 3-9.

2. A. Andreotti, On a Theorem of Torelli, Amer. J.Math. 80(1958), 801-828.

3. A. Andreotti and A. L. Mayer, On Period Relations for Abelian Integrals on Algebraic Curves, Annali Sc. Nor. Sup. Pisa, 21 (1967), 189-238.

4. W. L. Baily, Jr., On the Theory of θ-functions, the Moduli of Abelian Varieties and the Moduli of Curves, Ann. of Math. 75 (1962), 342-381.

5. L. Bers, Spaces of Riemann Surfaces, Proc. Int. Congr. Math., Cambridge, 1958, 349-361.

6. B. Christoffel, Vollständige Theorie der Riemann'schen θ-function. Math.Annalen, 58 (1901), 347-399.

7. Conforto, Abelsche Functionen und Algebraische Geometrie, Springer, Berlin, 1956.

8. T. Hayashida and M. Nishi, Existence of Curves of Genus Two on a Product of Two Elliptic Curves, Jl. Math. Soc. Japan, 17 (1965), 1-16.

9. W. L. Hoyt, On Products and Algebraic Families of Jacobian Varieties, Ann. of Math., 77 (1963), 415-423.

10. S. Lang, Abelian Varieties, Interscience, New York, 1959.

11. J. Lewittes, Riemann Surfaces and the Theta Function, Acta Math., 111 (1964), 37-61.

12. H. H. Martens, Classical Theory of Jacobian Varieties, Oslo University Math. Seminar (1964) No. 10.

13. H. H. Martens, On the Varieties of Special Divisors on a Curve, Crelles Jl., 227, (1967), 111-120.

14. T. Matsusaka, On a Characterization of a Jacobian Variety, Mem. Coll. Sci Kyoto, A, 32, No. 1, 1959.

15. A. Mayer, Special Divisors and the Jacobian Variety, Math. Annalen, 153 (1964), 163-167.

16. T. Meis, Die Minimale Blätterzahl der Konkretizierung einer Kompakten Riemannsche Fläche, Schr. Math. Inst. Münster (1960).

17. D. Mumford, On the Equations Defining Abelian Varieties, Inventiones Math., 1(1966), 287-354; 3(1967), 75-135; 3(1967), 215-244.

18. H. E. Rauch, A Transcendental View of the Space of Algebraic Riemann Surfaces, Bull. Amer. Math. Soc., 71 (1965), 1-39.

19. H. E. Rauch and H. M. Farkas, Relations between two Kinds of Theta Constants on a Riemann Surface, Proc. Nat. Acad. Sci. U.S.A., 59(1968), 52-55.

20. B. Riemann, Gesammelte Werke, 2nd Ed. Dover, New York, 1953. (Especially Articles VI, XI, XXXI, and Nachträge.)

21. F. Schottky, Zur Theorie der Abelschen Funktionen... Crelle's Jl., 102 (1888), 304-352.

22. F. Severi, Vorlesungen über Algebraische Geometrie, Teubner, Leipzig, 1921.

23. R. Torelli, Sulle Varietà di Jacobi, Rendiconti A. R. A. d. Lincei, 22 (1913), 98.

24. A. Weil, Variétés Abeliénnes et Courbes Algébriques, Hermann, Paris, 1948.

25. A. Weil, Zum Beweis des Torellischen Satzes, Nachr. Akad. Wiss. Göttingen, Math-Phys. Kl. II (1957), 33-53.

RECENT DEVELOPMENTS IN THE THEORY OF DISCONTINUOUS GROUPS OF MOTIONS OF SYMMETRIC SPACES

Atle Selberg

I shall in this lecture attempt to give some idea of recent advances as well as to comment on some open problems. The lecture concerns developments over the last decade or so, but as it is mainly intended for the nonspecialist, the introductory part giving the background is rather long. I have also not gone into much technical detail.

In as far as the important work of Borel and Harish-Chandra [1]*) on arithmetical groups has been omitted, since my emphasis has been in a quite different direction, this clearly does not pretend to be a complete account. It is also probably a biased one since I have selected the material and aspects of the subject that appeal to my own taste (or prejudice?).

1. Let S be a globally symmetric riemannian space, that is: for any point x in S there exists a global isometry μ_x which is an involution with the point x as an isolated fixpoint.**) We shall assume that S contains no compact symmetric space as a factor (this is equivalent to saying that μ_x should have no other fixpoint than x), and also that no euclidean factor is present in S. If S is not a product of symmetric spaces of lower dimension, we say that S is <u>irreducible</u>, otherwise S is said to be <u>reducible</u>. By the <u>rank</u> of S we understand the dimension of a maximal flat totally geodesic submanifold of S. We denote the connected component (identity component) of the group of isometries of S by G (consisting actually of the isometries composed by an even number of involutions μ). By K we denote a maximal compact subgroup of G, the space S can then be identified with the quotient- or coset-space G/K. G can be realized (in various ways) as a complex or real matrix group, that is to say a subgroup of either $S\ell(n,C)$ or $S\ell(n,R)$ for some n, where the entries g_{ij} or coefficients in the n by n matrix (g_{ij}) and their complex conjugates satisfy certain algebraic relations. The identification of G with the matrix group requires that we divide out the center of the matrix group, however we will in the following use G to denote the matrix group also in many cases, as no confusion is likely to arise.

The simplest example of such a space S is the hyperbolic plane of which a model is given

---

*) Numbers in brackets refer to the list of references at the end.

**) See for instance S. Helgason [6].

by the upper complex half plane $\mathcal{H}$, $z = x + iy$, $y > 0$ with the metric $ds^2 = \frac{dx^2 + dy^2}{y^2}$.
The group of motions G is then the group of linear transformations

$$z \longrightarrow \frac{az+b}{cz+d} \quad (1.1)$$

where $ad - bc = 1$, and a, b, c and d are real. It is also represented by the matrix-group $s\ell(2,R)$ with elements $\begin{pmatrix} a & b \\ c & d \end{pmatrix}$ if we identify $\begin{pmatrix} a & b \\ c & d \end{pmatrix}$ and $\begin{pmatrix} -a & -b \\ -c, & -d \end{pmatrix}$. As is well-known, the geodesics in this case are circles orthogonal to the real axis, and the transformations or motions (1.1) that are different from the identity, are classified as hyperbolic, elliptic and parabolic according as the transformation (1.1) has two real, two conjugate complex, or only one real fixpoint, or what is the same, according as $|a+d| > 2$, $|a+d| < 2$ or $|a+d| = 2$. By a suitable inner automorphism of G, these can be brought on the normal forms

$$\begin{pmatrix} a & 0 \\ 0 & \frac{1}{a} \end{pmatrix} \text{ with } a > 1, \quad \begin{pmatrix} \cos\frac{\alpha}{2}, & \sin\frac{\alpha}{2} \\ -\sin\frac{\alpha}{2}, & \cos\frac{\alpha}{2} \end{pmatrix}, \quad \text{and} \quad \begin{pmatrix} 1 & b \\ 0 & 1 \end{pmatrix}. \quad (1.2)$$

Other examples of symmetric spaces are given for instance by the space of positive definite symmetric n by n matrices Y with determinant 1 and the metric $ds^2 = \sigma(Y^{-1}dY\, Y^{-1}dY)$, σ denoting the trace. The corresponding group G is $s\ell(n,R)$. Similarly the space of positive definite Hermitian matrices is associated with the group $s\ell(n,C)$. Other matrix groups that correspond to symmetric spaces are the orthogonal (unitary) group $SO(p,q)$ $(SU(p,q))$ of an indefinite symmetric (hermitian) quadratic form of signature (p,q). The space corresponding to $SO(p,q)$ is the n-dimensional hyperbolic space. Other examples are C. L. Siegel's symplectic space and group, the bounded symmetric domains with the Bergmann metric. There is quite a bit of overlapping among these examples, $s\ell(2,C)$ for instance is also associated with the hyperbolic three-dimensional space.*)

A space that we shall have occasion to look at particularly in the following is $\mathcal{H}^n$, the product of n upper half planes or hyperbolic planes $\mathcal{H}$.

2. By a discontinuous group Γ of motions of S, we understand a discrete subgroup of G. We shall in the following denote the elements of G by g, and of Γ by γ (however, when we are looking at a representation of G as a matrix group we will use capital letters to denote elements of G or Γ). We call two points x_1 and x_2 equivalent under Γ if

---

*) For a systematic list see S. Helgason [6].

$x_2 = \gamma x_1$ for some γ in Γ. By a fundamental set of Γ in S we understand a set of points which contains one and only one representative of each equivalence class of points in S, if the set is sufficiently well-behaved to warrant the name of domain, we call it a <u>fundamental domain</u> of Γ in S and denote it by $\mathcal{D}$ or $\mathcal{D}_\Gamma$. One useful method for producing such a fundamental domain is to choose a point x_0 in S which is not a fixpoint for any element in Γ apart from e (we use e to denote the identity in G, for matrices we use E as the identity) and including in our fundamental domain $\mathcal{D}$ all points x with the property that $d(x,x_0) < d(x,\gamma x_0)$ for all $\gamma \neq e$ in Γ. Here $d(x,x_0)$ denotes the geodesic distance. By making some suitable convention about including parts of the boundary of this open domain, we can get a fundamental domain in the strict sense. Such a domain is called a <u>canonical fundamental domain</u> with x_0 as <u>base point</u>. The fundamental domain is a convenient and useful way of representing $\Gamma\backslash S$ the geometric entity that arises from S by identifying equivalent points.

Often it is more convenient to consider Γ as acting on G rather than on S, a set of representatives of the right cosets of Γ in G, which we denote by $\Gamma\backslash G$ is then a fundamental set of Γ in G. Since $S = G/K$ the two things are essentially the same. In particular since K is compact it is clear that if the volume of $\mathcal{D}$, $V(\mathcal{D})$ or $V(\Gamma\backslash S)$ is finite, then $m(\Gamma\backslash G)$, where m denotes the (bi-invariant) Haar measure on G, is also finite and vice versa. The class of groups Γ for which $m(\Gamma\backslash G) < \infty$ is for reasons which largely have to do with the study of functions on S or G which are invariant (or, change in a particularly simple way) under the action of Γ, a particularly important one. In what follows we assume always that Γ belongs to this class. There are two possibilities, either $\Gamma\backslash G$ or (what is the same) the fundamental domain is compact or it is not compact. In the first case there is a compact region in S so that every point x has an equivalent point in this region. In the second case it is clear that $\mathcal{D}$ must contain parts that stretch infinitely far out. If $\mathcal{D}$ is a canonical fundamental domain with x_0 as base point, it is easy to see that $\mathcal{D}$ must contain in its entirety at least one geodesic ray beginning at x_0. We shall refer to this simply as a <u>geodesic ray of</u> $\mathcal{D}$. The set of geodesic rays of $\mathcal{D}$ may consist of isolated rays or they may form continua. We shall refer loosely to the parts of $\mathcal{D}$ that lie in the neighbourhood of the far parts of these rays or continua of rays as the "non-compact parts" of $\mathcal{D}$ or the "cusps" of $\mathcal{D}$.

3. It will be useful to introduce certain concepts for the purpose of comparing and classifying different groups Γ. We say that two groups Γ and Γ' are <u>similar</u> if one is obtained from the other by an inner automorphism or conjugacy of G so that $\Gamma' = g\Gamma g^{-1}$ with g in G. If Γ_t is a group that depends continuously on a parameter t (meaning that the elements of Γ_t depend continuously on t),*) in say the interval $0 \leq t \leq 1$, we say that Γ_t represents a <u>deformation</u> of Γ, and we also call Γ_0 and Γ_1 <u>related</u> groups. If the deformation can be represented by $\Gamma_t = g_t\Gamma_0 g_t^{-1}$ where g_t is an element of G depending on t, we call the deformation <u>trivial</u>, otherwise we say it is <u>nontrivial</u>. If Γ does not admit any nontrivial deformation we call it <u>rigid</u>. The set of groups related to Γ if we identify those that are similar, we refer to as the <u>family</u> of Γ. If Γ is rigid, the family consists of just one member, if Γ is not rigid the family consists of groups depending in a nontrivial way continuously on a certain number of parameters or moduli.

If two groups Γ and Γ' have the property that their intersection is a subgroup of finite index in one of them (because of our assumptions it will then be so in both), we say that Γ and Γ' are strictly commensurable. Two groups Γ and Γ' are said to be <u>commensurable</u> if Γ is strictly commensurable to a group which is similar to Γ'. Commensurability is easily seen to be an equivalence relation.

If one wishes to describe and classify the groups Γ for a space S, one can attempt to do this on the basis of similarity, family, or finally, commensurability. What one does will depend somewhat on the nature of the groups for this particular space (are they all rigid for instance, family and similarity mean the same thing), and also how fine a classification one thinks one can accomplish (commensurability of course gives a cruder classification than similarity).

For euclidean n-dimensional space the corresponding problems are of course much easier, classification based on commensurability is quite simple to carry out for general n, based on similarity it can in principle be carried out for each n and for $n < 4$ a complete catalogue has been made.

---

*) In such a way that the structure, the discreteness and the finiteness of the volume of the fundamental domain is preserved.

4. We shall look briefly at what is known in the case of the hyperbolic plane, using the upper half plane model mentioned in section 1. Already Gauss considered the classical modular group Γ given by $S\ell(2,Z)$ the subgroup of $S\ell(2,R)$ where a, b, c and d in the elements (1.1) are rational integers. A fundamental domain $\mathcal{D}$ is arrived at, if we use for instance the point $2i$ as base point, which is the region $|z| > 1$, $|x| < \frac{1}{2}$ with certain parts of the boundary included. The fundamental domain is not compact, and $\mathcal{D}$ contains one geodesic ray from the base point. The noncompact part or cusp of $\mathcal{D}$ is the upper part of the strip $|x| < \frac{1}{2}$ and is determined or described fully by a subgroup of Γ, the group $\begin{pmatrix} 1 & m \\ 0 & 1 \end{pmatrix}$ where m runs over the integers. This is a group of parabolic transformations according to the classification in section 1, which leave the point at infinity, the endpoint of the geodesic ray fixed. This group is also the first example of what we call an arithmetical group.*)

Later other groups were constructed by starting with a geodesic polygon with a finite number of sides, whose interior angles were of the form $\frac{\pi}{q}$ where the q's are integers > 1, or $q = \infty$ if the vertex is on the real axis or at infinity. By reflecting over the sides of this polygon one generates a group. The subgroup of index two whose elements are composed of an even number of reflections is then a discrete group of motions and the original polygon plus a reflected image of it is a (not necessarily canonical) fundamental domain for this group. If some interior angles are zero the fundamental domain is noncompact and has cusps like that of the modular group.

Finally the uniformization theory as developed by Poincaré, Klein and Koebe showed that the conformal mapping of certain covering surfaces of algebraic riemann surfaces into the upper half plane gave rise to groups Γ in $\mathcal{H}$ which correspond to the transformations of the covering surface into itself. This approach actually gave all groups with $V(\Gamma\backslash\mathcal{H}) < \infty$, and a complete description of them in terms of generators and relations.**) They are all finitely generated with a finite number of independent relations. A canonical fundamental domain is always a polygon with a finite number of sides. While there are some groups that are rigid (the so-called triangle groups obtained by the reflection of a triangle with angles $\frac{\pi}{p}, \frac{\pi}{q}, \frac{\pi}{r}$ with $\frac{1}{p} + \frac{1}{q} + \frac{1}{r} < 1$, where p, q, r are positive integers or ∞, the modular group is one of these), the others form families of related groups that depend analytically on a

---

*) For a general definition of this see [1].

**) See for instance [4].

certain number of parameters or "moduli", this number can be arbitrarily large for suitable families.

In case $\Gamma\backslash\mathcal{H}$ is not compact, the only type of noncompactness that is found is of the type found in the modular group, that is: cusps that are determined by a subgroup of parabolic or unipotent matrices (matrices that can be brought by a conjugacy on triangular form with 1's on the main diagonal and zeros below it).

In the hyperbolic plane there is far greater variety of groups Γ and far more freedom in producing groups with certain arbitrarily assigned properties than there is in the case of the corresponding situation in the euclidean n-dimensional space. In particular, the topological characteristic, the genus, of $\Gamma\backslash\mathcal{H}$ (which is, apart from some possible exceptional points, essentially a closed orientable surface) can be arbitrarily prescribed.

5. In the case when the dimension of S is > 2, very little general information had been obtained until recently. Essentially the only groups known in these cases were arithmetical groups defined in the manner of the modular group as a subgroup of the matrix group G determined by some kind of integrality condition. Such is for instance the Hilbert modular group acting on the space $\mathcal{H}^n$. G in this case is $(S\ell(2,R))^n$. If k is a totally real number field of degree n over the rationals, and $\theta^{(i)}$ $i = 1,2,\ldots,n$, denotes the n conjugates of θ when θ belongs to the field, we consider integers α, β, γ and δ in k such that $\alpha\delta - \beta\gamma = 1$. The matrices

$$\begin{pmatrix} \alpha^{(i)} & \beta^{(i)} \\ \gamma^{(i)} & \delta^{(i)} \end{pmatrix} \qquad i = 1,2,\ldots,n \tag{5.1}$$

are then the n components of an element in the Hilbert modular group of the field k.

Other examples of arithmetical groups were $S\ell(n,Z)$, the unit groups of indefinite quadratic forms with rational coefficients, the symplectic modular group and several other general examples which had been investigated notably by C. L. Siegel.

For the three-dimensional hyperbolic space there existed one or two examples of groups constructed by reflections of simple polyhedra.

Nothing like the uniformization theory was available in higher dimensions.

Nothing was really known about the structure and other properties of groups Γ that were implied by the condition $V(\Gamma\backslash S) < \infty$, whether such groups were rigid or whether families depending on "moduli" existed.

If $\Gamma\backslash S$ was noncompact it was not known whether Γ had a finite number of generators with a finite number of relations or not, whether a canonical fundamental domain had to be a polyhedron bounded by a finite number of hypersurfaces. The nature of the noncompact parts or cusps of the fundamental domain and of the subgroups of Γ that determined or described them was also unknown.

It was natural to think that in general things would be much as one had found them for the arithmetical groups that had been investigated in detail. The fact that no groups were known that were not commensurable with some arithmetical group (the examples constructed by reflection for the three-dimensional hyperbolic space had not been investigated with this in mind), might also lead one to speculate on the relationship between the arithmetical groups and all groups with $V(\Gamma\backslash S) < \infty$, when the dimension of S was > 2. It was not at all obvious what to expect.

This was the state of affairs a little more than ten years ago. Today a fairly significant beginning has been made. Some questions have been answered, and for some we have partial answers. In some cases apparently natural first guesses have had to be considerably modified.

6. In order that we, in the following, shall be able to state results in a simple form also when the space S is reducible, we need to introduce certain concepts that serve to distinguish groups Γ that "properly" belong to S from those that really have to do with some factorization of S. When S is reducible G is of course a direct product of the groups of motions of the irreducible factors of S.

We now call Γ reducible if there is a factorization of S in S_1 and S_2 such that the restriction of Γ to S_1 is a discrete subgroup of G_1. One can show that then the restriction of Γ to S_2 is also a discrete subgroup of G_2, and that Γ is commensurable to a group which is the direct product of a discrete subgroup Γ_1 of G_1 and another discrete subgroup Γ_2 of G_2.

A group Γ which is not reducible we call irreducible (in particular if S is irreducible Γ is too).

If Γ is irreducible and S_1 is a proper factor of S, one can show that the following two statements hold.*)

*) Selberg [12], p. 163. In the case $S = \mathcal{H}^n$ also found independently by Piatetsky-Shapiro [11].

Let Γ_1 denote the restriction of Γ to S_1, then

(1) Only the identity element of Γ corresponds to the identity element of Γ_1,

(2) Γ_1 is dense in G_1 the group of motions of S_1.

The Hilbert modular group defined in the preceding section is an example of an irreducible group Γ acting on a reducible space.

7. We shall in this section look in some detail at the situation for the space $\mathcal{H}^n$. Apart from being of considerable independent interest, this will serve later as a proper background against which to present the current state of affairs for general S.

7.1. While in the general theory there has always been a sharp distinction between the case when $\Gamma\backslash S$ is compact and when it is not, it is a curious fact that the very first result obtained about groups in general for a symmetric space S of dimension > 2, was obtained without making any such distinction.

We assume Γ to be an irreducible group acting on $\mathcal{H}^n$, and that we have given a deformation of Γ. We fix our attention on one of the factors $\mathcal{H}$ and denote by γ_1 the restriction of any element γ in Γ to this factor. Using the statements (1) and (2) from the previous section,*) we see that the γ_1 lie everywhere dense in the set of all elliptic elements of $\mathcal{SL}(2,R)$, and further that during the deformation the trace of γ_1 or the rotation angle of the rotation of $\mathcal{H}$ that γ_1 signifies when it is elliptic has to remain constant during the deformation. Namely if the rotation angle varies, it would have to change from an irrational multiple of π to a rational multiple of π or vice versa, which would mean that γ_1 changes from an element of infinite order to one of finite order or vice versa. Because of (1) of section 6, if γ_1 is of finite order so is γ, thus γ itself would change from infinite to finite order (or the other way around). This is of course incompatible with the preservation of the structure of Γ. If we consider those γ_1 which are elliptic and of infinite order, it is easy to see that these are still dense in the set of all elliptic elements of $\mathcal{SL}(2,R)$. If γ_1 and γ_1' both are elliptic and one of them, say γ_1' is of infinite order, then it is easy to see that we can find positive integers m such that $\gamma_1'' = \gamma_1\gamma_1'^m$ is again elliptic. The fact that the trace of γ_1'' also must remain constant is

*) At the time the proof which is sketched here was found (1957), I did not have the general form of (1) and (2), but only the case of $\mathcal{H}^n$, when they can be simply proved, using the list given by J. Nielsen of all closed subgroups of $\mathcal{SL}(2,R)$ which are not discrete.

easily seen to imply that the geodesic distance between the fixpoints of γ_1 and γ_1' also remains constant during the deformation. This is seen to imply that the deformation of the set of fixpoints of all elliptic γ_1 is an isometry, actually a motion of $\mathcal{H}$. Thus the set of all γ_1 which are elliptic only undergoes an inner automorphism of $S\ell(2,R)$ by the deformation of Γ. Since furthermore any γ_1 can be written as a product of two elliptic elements γ_1' and γ_1'', it follows that the full set of γ_1's just undergoes some inner automorphism of $S\ell(2,R)$. Arguing in the same way for the other factors $\mathcal{H}$, we conclude that the deformation of Γ is an inner automorphism of $(S\ell(2,R))^n$ and therefore a trivial deformation. Thus Γ is rigid.

7.2. The Hilbert modular group introduced in section 5, has a cusp which is described by the subgroup formed by the conjugates of the matrices

$$\begin{pmatrix} \eta & \omega \\ 0 & \eta^{-1} \end{pmatrix} \tag{7.2.1}$$

where η runs over the totally positive units and ω over the integers of the field. A very important step in the investigation of irreducible groups with noncompact $\Gamma\backslash\mathcal{H}^n$ was made by Piatetsky-Shapiro [11], when he was able to show that if Γ was irreducible and possessed a unipotent element, that is one of which all components were parabolic, then this element belonged to a subgroup of Γ which described a cusp, and furthermore this subgroup had to be commensurable with the group (7.2.1) for some totally real numberfield of degree n over the rationals, that is: commensurable to the group describing the cusp at "infinity" for some Hilbert modular group.

His proof was based on the simple fact that in a sufficiently small neighbourhood of the identity e, the group G is very nearly abelian. Thus, if one introduces some measure of distance between two elements $d(q_1,q_2)$, either by making G into a riemannian space by introducing a positive definite metric invariant under multiplication on the left as can be done, or by using a matrix representation of the group G and $\sigma((\overline{A-B})^t(A-B))^{\frac{1}{2}}$ as the measure of the distance between two matrices A and B, it will be true that if Ω is a sufficiently small open neighbourhood of e, and q_1 and q_2 lie in Ω then the commutator $[q_1,q_2] = q_1q_2q_1^{-1}q_2^{-1}$ has the property

$$d([q_1,q_2],e) \leq \frac{1}{2}\min(d(q_1,e),d(q_2,e)). \tag{7.2.2}$$

Thus if we form successive commutators

$$[q_1,q_2],\ [q_1,[q_1,q_2]],\ \dots,[q_1,[q_1,\cdots [q_1,q_2]]\cdots],\dots$$

this sequence must converge to e.

If Γ is a discrete subgroup of G and ρ and γ two elements of Γ that lie in Ω then the sequence

$$(7.2.3) \qquad \gamma_1 = [\rho,\gamma],\ \cdots\ \gamma_m = [\rho,\gamma_{m-1}],\ \cdots$$

must have all terms equal to e after a finite number of steps. The same is easily seen to be true if γ lies in the larger region $G_\rho \cap G_\rho$ where G_ρ denotes the centralizer of ρ in G. If Ω' is a smaller open neighbourhood of e such that $\Omega' \Omega'^{-1}$ is contained in Ω, then any γ in Γ which has the property that $\gamma G_\rho \Omega'$ and $G_\rho \Omega'$ has a nonempty intersection, must lie in $G_\rho \cap G_\rho$. If G_ρ is not compact the Haar measure of the set $G_\rho \Omega'$ in G is infinite, so that there must exist (since $m(\Gamma\backslash G) < \infty$) enough γ's in $G_\rho \cap G_\rho$ to make $m(\Gamma\backslash G_\rho \Omega') < \infty$. The fact that the sequence (7.2.3) equals e after a finite number of terms, then enables us to say something about these γ, and to deduce that Γ possesses a subgroup of a certain kind. Sometimes it is possible to repeat similar arguments based on commutator formation, using other elements from this subgroup also and so obtain the existence of a still larger subgroup.

The general type of argument sketched above, was used by Piatetsky-Shapiro with the unipotent element of Γ whose existence he postulated, as the ρ above, since by a suitable inner automorphism of G a unipotent element can always be brought into a prescribed small open neighbourhood of e. The arguments clearly apply, and give in this case first that the group $\Gamma_\rho = \Gamma \cap G_\rho$ has the property that $m(\Gamma_\rho \backslash G_\rho)$ is finite in the Haar measure on G_ρ. In the next step we show that if we denote by G_ρ^* the subgroup of G which has the property that if g is in G_ρ^*, the operation $g\, G_\rho\, g^{-1}$ not only carries G_ρ into itself, but also leaves the Haar measure on G_ρ invariant, then the subgroup of Γ, $\Gamma_\rho^* = \Gamma \cap G_\rho^*$ has the property that

$$m(\Gamma_\rho^* \backslash G_\rho^*) < \infty$$

measured in the invariant Haar measure on G_ρ^*. From this it follows simply that Γ has a subgroup commensurable with that describing the cusp at "infinity" for some Hilbert modular group.

It is the fact that G_ρ^* is so "large" that the quotient $G_\rho \backslash G_\rho^*$ with its (normal) subgroup of unipotent elements G_ρ is not compact, which brings in the algebraic number field and the arithmetical structure in Γ_ρ^*. The rank of the space $\mathcal{H}^n$ is n, and this an example of an essential difference between the situation for rank > 1 and for rank $= 1$.

7.3. Holding still to the hypothesis that the irreducible subgroup Γ of $(S\ell(2,R))^n$ has a unipotent element, one can go further and draw conclusions about the whole group Γ. If ρ is the unipotent element, we can by an inner automorphism of G make it so that Γ_ρ^* is strictly commensurable with the subgroup of the Hilbert modular group that describes the cusp at infinity. Because of the "largeness" of G_ρ^* referred to above, it can be seen that if ρ_1 is another unipotent element of Γ which does not lie in Γ_ρ, then the two groups Γ_ρ^* and $\Gamma_{\rho_1}^*$ must have an infinite subgroup in common. Actually in this case $G_\rho^* \cap G_{\rho_1}^*$ is not compact and $\Gamma_\rho^* \cap \Gamma_{\rho_1}^*$ can be shown to be a subgroup whose quotient has finite measure (actually the quotient is compact). In particular this means that the algebraic number field connected with Γ_ρ^* is the <u>same</u> for all ρ in Γ and that all groups Γ_ρ^* are commensurable. By comparing Γ_ρ^* with $\Gamma_{\gamma\rho\gamma^{-1}}^*$ where γ is an arbitrary element in Γ but not in Γ_ρ^*, we find that there must exist some relation

$$\sigma_1\gamma = \gamma\sigma_2$$

where σ_1 and σ_2 are elements of $\Gamma_\rho^* \neq e$. Similar relations exist for the powers of $\gamma \neq e$. From this it is possible to conclude that all elements of Γ are of the form

$$\frac{1}{\sqrt{\Delta^{(i)}}}\begin{pmatrix}\alpha^{(i)} & \beta^{(i)}\\ \gamma^{(i)} & \delta^{(i)}\end{pmatrix} \qquad (i = 1,2,\ldots,n) \tag{7.3.1}$$

where α, β, γ and δ are integers in the totally real number field, $\Delta = \alpha\delta - \beta\gamma$ is totally positive, and the $\alpha^{(i)}$ denotes the (i)th conjugate of α.

7.4. As we have seen one can get very much out of assuming the existence of one unipotent element. However, one would like to get rid of this assumption, and assume only that Γ is irreducible and $\Gamma\backslash\mathcal{H}^n$ not compact (and of course the everywhere implied assumption of the finite volume of $\Gamma\backslash S$). So it becomes necessary to prove the existence of a unipotent element from the noncompactness. (That unipotent elements or in general nondiagonalizable matrices do not occur when $\Gamma\backslash S$ is compact is almost trivial.) If we turn our attention to a geodesic ray of a canonical fundamental domain of Γ with x_0 as base point, this ray can be described by $x_t = g^t x_0$, $t \geq 0$, where g^t belongs to a one-parameter subgroup of G (actually obtained by forming $\mu_x \mu_{x_0}$, where x is any point on the ray and μ_x denotes the involution mentioned in section 1). If we could show that there is an element $\rho \neq e$ in Γ such that as $t \longrightarrow \infty$

$g^{-t}\rho g^{t} \longrightarrow e$, then ρ must be a unipotent element which leaves the infinite point on the ray fixed. It is easy to show that there is a sequence of elements in Γ $\rho_1, \rho_2, \ldots, \rho_m, \ldots \neq e$, such that the sequence $g^{-m}\rho_m g^m \longrightarrow e$, and, by using arguments based on the formation of commutators similar to those sketched in 7.2 above, it is also easy to make it seem reasonable that the ρ_m from a certain m of have to be unipotent, and if we use certain minimal conditions to make the choice of ρ_m unique, that the ρ_m from a certain m of have to be the <u>same</u> unipotent element. It is not quite so simple to prove this to be true, however. Using certain of the specific properties of $S\ell(2,R)$, (mainly that the centralizer of an element $\neq e$, always is abelian), I succeeded in 1964 in proving the statement that there exists a unipotent element ρ with $\lim_{t\to\infty} g^{-t}\rho g^{t} = e$ for the case of $\mathcal{H}^n$. The proof as originally presented was rather long and cumbersome.

Using this, together with the material of 7.2 and 7.3, one can show: Γ has a canonical fundamental domain which has a finite number of cusps or noncompact parts which are described by subgroups of Γ of the type discussed in 7.2 and no other kind of cusps. The set of geodesic rays of the fundamental domain is finite with one ray for each cusp. The fundamental domain is a polyhedron bounded by a finite number of hyperplanes. Γ has a finite set of generators with a finite set of independent relations. If Γ is brought by an inner automorphism of G on the form given in 7.3 where all elements are of the form (7.3.1), then for every subgroup describing a cusp of the corresponding Hilbert modular group, Γ has a subgroup describing a cusp so that the two are strictly commensurable.

7.5. The preceding results make it natural to speculate on whether Γ itself is necessarily commensurable to the corresponding Hilbert modular group.

This would actually follow if we could show that Γ is a group with <u>bounded denominator</u>. By this we mean that by multiplying the matrices representing the elements in Γ with some fixed rational integer, we could make all entries or matrix coefficients become integral algebraic numbers. It is easy to see that a matrix group whose elements have entries that all lie in some algebraic number field, and which has bounded denominator, necessarily possesses a subgroup of finite index which has only integral algebraic numbers in the matrices. In the present case, since Γ is finitely generated, the subgroup of Γ which has as elements those that can be written with each generator occurring an even number of times (counted with their exponents as weights), will be of finite index in Γ and will have all the $\Delta = \alpha\delta - \beta\gamma$ equal to squares of numbers in the field. Thus the entries in the matrices

$$\frac{1}{\sqrt{\Delta}}\begin{pmatrix}\alpha & \beta \\ \gamma & \delta\end{pmatrix}$$

become numbers in the same number field that α, β, γ and δ belongs to for this subgroup. This subgroup again, if the denominator is bounded, has a subgroup of finite index where the entries in the matrices are integral. This last subgroup is a subgroup of the Hilbert modular group and so Γ is commensurable with this.

We can attack the problem of whether Γ has bounded denominator in the following way:

It is possible to bring Γ to the form (7.3.1) in such a way that it contains the subgroup consisting of the elements

$$S_\omega = \begin{pmatrix}1 & \omega \\ 0 & 1\end{pmatrix} \tag{7.5.1}$$

where ω runs over all the integers of the algebraic number field. We may assume (by eventually passing to a subgroup of finite index), that all elements of Γ are of the form

$$M = \begin{pmatrix}\alpha & \beta \\ \gamma & \delta\end{pmatrix} \tag{7.5.2}$$

where α, β, γ, δ are in the field and $\alpha\delta - \beta\gamma = 1$. If we can show that all the γ's are necessarily integral, we could, by utilizing the presence in Γ of an element of the form

$$R = \begin{pmatrix}1 & 0 \\ \omega_0 & 1\end{pmatrix}$$

where ω_0 is an integer in the field $\neq 0$ (this follows from the results in 7.4), and looking at RM and MR conclude that $\omega_0\,\alpha$ and $\omega_0\,\delta$ always are integral and also $\omega_0^2\,\beta$, from this would follow that Γ has bounded denominator.

We therefore now assume that there is an element of the form (7.5.2) where γ is <u>not</u> integral, and we seek to arrive at a contradiction. We write

$$M = S_{\frac{\alpha}{\gamma}}\begin{pmatrix}0 & -\frac{1}{\gamma} \\ \gamma & 0\end{pmatrix} S_{\frac{\delta}{\gamma}} = S_{\frac{\alpha}{\gamma}}\, T_\gamma\, S_{\frac{\delta}{\gamma}} \tag{7.5.3}$$

and consider expressions of the form

$$M\,S_{\omega_1}\,M^{-1}\,S_{\omega_2}\,M\,S_{\omega_3}\cdots S_{\omega_{m-1}}\,M^{(-1)^{m-1}}\,S_{\omega_m}\quad . \tag{7.5.4}$$

This is seen to differ (since all S commute) only in a very simple way from the expression

(7.5.5) $$T_\gamma S_{\omega_1} T_\gamma S_{\omega_2} \cdots S_{\omega_{m-1}} T_\gamma S_{\omega_m}$$

which is rather simple to compute recursively. Apart from the factor $(-1)^{\left[\frac{m}{2}\right]}$ (7.5.4) and (7.5.5) have the same number in the lower left corner of the matrix, and when m is even they have the same trace. We seek to produce an element of Γ of the form (7.5.4) which has eigenvalues in the field, but such that these are not units of the field, this can be seen to contradict what we stated in section (7.4) about the subgroups of Γ describing the cusps. If the ideal $(\gamma) = \frac{\mathfrak{z}}{\mathfrak{q}}$, where $\mathfrak{z}$ and $\mathfrak{q}$ are relatively prime ideals and we look at (7.5.4) for $m = 2$ we find its trace to be $2 - \gamma^2 \omega_1 \omega_2$, or writing ω for $\omega_1 \omega_2$ $2 - \gamma^2 \omega$, if σ is an integral number, not a unit, such that $(\sigma) | \mathfrak{q}^2$, then we can show that we may choose ω such that one eigenvalue is of the form $\frac{\eta}{\sigma}$ with η a unit, if the congruence

(7.5.6) $$\sigma \equiv \eta (\mathfrak{z})$$

is satisfied. Though there is some freedom here in the choice of a divisor σ of $\mathfrak{q}^2$ and in the choice of the unit η, this congruence may not have a solution. What we try is to produce from the form (7.5.4), an element with a γ such that (7.5.6) has a solution. In (7.5.4) the new γ^* will in general have the form

$$(\gamma^*) = \frac{\mathfrak{z}^*}{\mathfrak{q}^*}$$

where $\mathfrak{q}^* = \mathfrak{q}^m$, and $\mathfrak{z}^*$ contains the factor $\mathfrak{z}$ or $\mathfrak{z}^2$ according as m is odd or even. The other factor of $\mathfrak{z}^*$ we seek to control through a suitable choice of the ω's, actually we try to make it a power of a prime ideal whose norm is a rational prime and which satisfies certain other conditions. What remains is a purely number theoretical problem, which I unfortunately have not been able to solve. However, it is possible to prove that we can produce a new γ for which (7.5.6) is solvable, if we assume certain hypotheses.

One such which shall be mentioned here, because it probably will seem a most plausible hypothesis to most mathematicians, is the following.

Associated with the number field of Γ, we can define L-functions, $L(s,\chi)$ much like the dirichlet L-functions, when we have a character of a residue system for modulo an ideal $\mathfrak{m}$; in order that this character should be extendable to ideals it is necessary and sufficient that associated numbers have the same character (or otherwise expressed that $\chi(\eta) = 1$ for all units η). If this is so then

$$L(s,\chi) = \Sigma \frac{\chi(\mathfrak{m})}{N(\mathfrak{m})^s}$$

where the sum extends over all integral ideals, is a function with an Euler product, if χ is not the principal character it is an integral function, with a functional equation relating $L(s,\chi)$ and $L(1-s,\bar{\chi})$. If we make the assumption that there is an open neighbourhood of $s = 1$ depending only on the field such that no $L(s,\chi)$ vanishes in this neighbourhood, then our construction of a γ for which (7.5.6) is solvable works; so it follows that Γ is then commensurable to the Hilbert modular group.

7.6. It should be pointed out that the developments in 7.2 to 7.5 can be carried through without any real change in a somewhat more general context, namely where the space S is a product of r_1 hyperbolic planes and r_2 hyperbolic three-dimensional spaces and $r_1 + r_2 > 1$. The corresponding group $G = (S\ell(2,R))^{r_1} (S\ell(2,C))^{r_2}$.

In place of the Hilbert modular group comes an arithmetical group associated with a number field of degree $r_1 + 2r_2$, which has r_1 real and $2r_2$ complex conjugates, in the same way that the Hilbert modular group is associated with the totally real number field. Here the real conjugates give the components from $S\ell(2,R)$ and one of each pair of complex conjugates the components from $S\ell(2,C)$.

Again we can show if we assume the existence of an open neighbourhood of $s = 1$, depending only on the field, such that $L(s,\chi) \neq 0$ in this neighbourhood; that any irreducible group with noncompact $\Gamma\backslash S$ but $V(\Gamma\backslash S) < \infty$ must be commensurable with the above mentioned arithmetical group.

8. We now turn to the case of general S assuming that $\Gamma\backslash S$ is compact and look at the question of rigidity. There have been essentially three different approaches to this problem.

One was a quite elementary approach working directly with some matrix representation of the group G, working essentially with the centralizers G_γ and $\Gamma_\gamma = \Gamma \cap G_\gamma$ of elements γ in Γ, and depending on the fact that the quotient $\Gamma_\gamma\backslash G_\gamma$ is compact. It may be possible to show that under a deformation of Γ certain quantities relating to Γ_γ have to remain invariant. Then by combining this information for the different γ one might be able to show that the deformation is necessarily trivial. This worked well for several general classes of symmetric spaces.*)

---

*) See Selberg [12] where the case $G = S\ell(n,R)$ for $n > 2$ was carried out in detail. Other cases that could be handled are listed there.

A quite different method which did not look directly at the group Γ and its elements, but rather on the space $\Gamma\backslash S$, using methods of differential geometry to investigate the possibility of varying the differential structure while in such a way that the metric still made $\Gamma\backslash S$ a locally symmetric space, was introduced by E. Calabi [2] at first only for the hyperbolic n-dimensional space, proving rigidity for $n > 2$. Later Calabi and Vesentini [3] proved rigidity in the case of the bounded symmetric complex domains, with of course the appropriate exception that the complex dimension be > 1 (to exclude the domain equivalent to the hyperbolic plane) and that if one-dimensional factors are present, Γ should be irreducible. With the exception of the domains of rank one, and those that involve any of the two exceptional domains, this could also be obtained by the elementary approach spoken of earlier.

Finally A. Weil [13] developing further the differential geometric method of Calabi gave a proof of rigidity for all irreducible groups Γ with compact $\Gamma\backslash S$ if S is not the hyperbolic plane. He also showed, what had been suspected for some time, that it was the set of relations between the generators of the group that kept it rigid.

Quite recently G. D. Mostow [10] in a note gave a new proof of rigidity for the case of the n-dimensional hyperbolic space when $n > 2$. His proof is quite simple, but it is not entirely clear how well the method which uses quasiconformal mappings of S onto itself, and looks at the extension of this mapping to the "boundary" of S, will carry over to the general case.

It should be mentioned that from rigidity of Γ and the fact that the rigidity is produced by relations of the (finite) set of generators, one can conclude in general the following:

If G is realized as a matrix group, such a group Γ is similar to one in which all numbers in the matrices are algebraic numbers from some algebraic number field.

9. When $\Gamma\backslash S$ is not compact, it was, apart from the result given in 7.1 earlier, much harder to obtain results, unless one made some kind of additional assumption beyond $V(\Gamma\backslash S) < \infty$. For instance, my proof of rigidity if $G = S\ell(n,R)$ with $n > 2$, could be carried over to the noncompact case, if one assumed that a canonical fundamental domain $\mathcal{D}$ was a polyhedron bounded by a finite number of hypersurfaces. This is of course a very strong assumption to make. Basically it is an assumption about the nature of the noncompact parts or cusps of $\mathcal{D}$, and also an assumption that there are only a finite number of such cusps. Results were also obtained by others under similar strong assumptions. Frankly, such results are not terribly interesting. It was clear that one had first to find out what the condition $V(\Gamma\backslash S) < \infty$ implied about the

cusps and the structure of the subgroups of Γ that described these cusps.

9.1. The first step necessary seemed to be to prove that for every geodesic ray of a canonical fundamental domain, there was some unipotent element of Γ leaving the infinite point on this geodesic ray fixed. In section 7.4 we saw that this could be proved if S is a product of hyperbolic planes, and in section 7.6 it was remarked that the proof could be carried out for products of r_1 hyperbolic planes and r_2 three-dimensional hyperbolic spaces, the restriction $r_1 + r_2 > 1$ mentioned there does not apply to this particular proof, so that the case of one hyperbolic three-dimensional space in particular is covered. I was later able to extend the proof to cover a few more cases, and Piatetsky-Shapiro and Vinberg (unpublished) also succeeded for some of the lower dimensional spaces. Quite recently D. A. Kazdan and H. A. Margulies [8] managed to give a general proof of this.

9.2. The next step would be to proceed somewhat in the manner indicated in 7.2 to get from the unipotent element as much information as possible about the rest of the subgroup describing the cusp this geodesic ray belongs to.

The simplest case is when the space has rank one. In this case I was able for the case of the n-dimensional hyperbolic space, and for the bounded symmetric domain $|z_1|^2 + \cdots + |z_n|^2 < 1$, where G is $SO(n,1)$ and $SU(n,1)$ respectively, to determine the structure of the cusps and the subgroups of Γ that describe them. I did not look at the other cases of rank one since at the time I was unaware of their existence. However, recently H. Garland and M. S. Ragunathan [5] by a treatment which differs in certain aspects, succeeded in obtaining the same results for all cases of rank one.*)

With this it is a simple thing to describe the canonical fundamental domain of such a group. It is of course a polyhedron bounded by a finite number of hypersurfaces, Γ has a finite number of generators and relations. The group that describes a cusp in this case always has a subgroup of finite index which consists only of unipotent elements apart from e. These subgroups are not rigid by themselves, unlike those occurring in the groups discussed in section 7.2 to 7.6, and have too little structure to bring anything of an arithmetical nature into the picture. However, it can be proved that the groups Γ are rigid**) if the dimension

*) At the time they did not know the details of my work; it seems according to communication from H. Garland, that my treatment of $SU(n,1)$ would have worked for the remaining cases of rank one also.

**) Carried out by H. Garland and M. S. Raghunathan [4] for all cases of rank one. For the hyperbolic spaces the argument of Mostow [10] should work well with a little adaptation.

of S is > 2. Also one can show that they, by an inner automorphism of G, can be brought on such a form as to have only numbers from some algebraic number field occurring in the matrices in Γ.

An example of a group in the hyperbolic 3-dimensional space shows that it is not necessarily the group relations that keep these groups rigid. The example, studied by J-P. Serre (oral communication), is such that the group relations permit deformations of the generators, but any nontrivial deformation destroys the discreteness of the group.

In recent years several examples of groups with $\Gamma\backslash S$ noncompact have been constructed for the 3-dimensional hyperbolic space, by means of reflections by Maharov, Vinberg and D. N. Verma. Some of these examples show the following things can happen:

(a) Γ need not be commensurable to an arithmetical group.

(b) The subgroups of Γ that describe two different cusps, need not be commensurable.

(c) In the complex lattice, that describes the group of unipotent elements belonging to a cusp, the "period ratio" or ratio between two independent elements (which is always an algebraic number) can be an algebraic number of arbitrarily high degree over the rational field.

9.3. In the case when the rank is > 1 it is more difficult to proceed from a unipotent element to a full description of a cusp. One reason is that in general, in contrast with the situation in the previous section and with that met for the groups

$$(S\ell(2,R))^{r_1}\ (S\ell(2,C))^{r_2}$$

earlier, there turns out to be quite many case distinctions to make. This is not surprising, since it is known that for certain spaces S there exist different arithmetical groups whose cusps are very essentially different, for instance such that the sets of geodesic rays associated with the cusps are of different dimension. One hopes of course eventually to show that in all cases of rank > 1, the only cusps that occur are described by subgroups of Γ that are commensurable with some subgroup of some arithmetical group describing a cusp of this.

If we for example consider the group $G = S\ell(3,R)$, we can show that if $\Gamma\backslash G$ is noncompact and of finite measure, there are two essentially different possibilities:

a. If the set of geodesic rays of a canonical fundamental domain is not finite, the group Γ can, by an inner automorphism of G, be brought on such a form that it has two subgroups, one formed by elements of the form

$$(9.3.1)\qquad \begin{pmatrix} a_{11} & a_{12} & a_{13} \\ a_{21} & a_{22} & a_{23} \\ 0 & 0 & 1 \end{pmatrix},$$

and another by elements of the form

$$(9.3.1')\qquad \begin{pmatrix} 1 & a_{12} & a_{13} \\ 0 & a_{22} & a_{23} \\ 0 & a_{32} & a_{33} \end{pmatrix};$$

where in each case the a_{ij} are rational integers, both of these groups are subgroups of finite index in the corresponding subgroup of $s\ell(3,Z)$.

b. If the set of geodesic rays is finite, then Γ can by an inner automorphism be brought on such a form that it possesses a subgroup of the form

$$(9.3.2)\qquad \begin{pmatrix} \eta & \alpha & \eta\beta \\ 0 & \tilde{\eta}^2 & \tilde{\eta}\tilde{\alpha} \\ 0 & 0 & \eta \end{pmatrix},$$

where α, β and η are integers in a real quadratic number field k, $\sim$ denotes conjugation with respect to the field and

$$\eta\tilde{\eta} = 1, \ \beta + \tilde{\beta} = \alpha\tilde{\alpha} .$$

This subgroup of Γ is of finite index in the group formed by taking <u>all</u> matrices of the form (9.3.2) where α, β and η satisfy these conditions. The latter group is easily seen to be one describing a cusp for the arithmetical group defined by the condition

$$(9.3.3)\qquad M \begin{pmatrix} 0 & 0 & 1 \\ 0 & -1 & 0 \\ 1 & 0 & 0 \end{pmatrix} \tilde{M}^t = \begin{pmatrix} 0 & 0 & 1 \\ 0 & -1 & 0 \\ 1 & 0 & 0 \end{pmatrix}$$

where the matrix M has entries that are integers in k and $\tilde{M}$ denotes the conjugate with respect to k.

It is possible to proceed from here in the manner of section 7.3, and to show that in case a,

Γ is similar to a group where all matrices are of the form

(9.3.4) $$\frac{1}{\sqrt[3]{\Delta}} \begin{pmatrix} a_{11} & a_{12} & a_{13} \\ a_{21} & a_{22} & a_{23} \\ a_{31} & a_{32} & a_{33} \end{pmatrix} ,$$

where the a_{ij} are rational integers, and Δ denotes the determinant $|a_{ij}|$. If we could show that Γ has bounded denominator (as defined in 7.5), it would obviously follow that in this case Γ is commensurable to the group $S\ell(3,Z)$.

It might seem promising to attack the question of the bounded denominator in the manner of section 7.5, since we have the two rather large subgroups at our disposal. However, there are certain complications in forming longer combinations of elements similar to the form (7.5.4).

In the case b, we could again show that Γ is similar to a group of matrices of the form (9.3.4) where now the a_{ij} would be integers in the real quadratic field k.

In both cases the group Γ has to be rigid and it is again so that this rigidity is imposed on the whole group by the subgroups describing a cusp. One can also show that the groups are finitely generated, in particular there is a finite number of inequivalent cusps, these have all to be of the same kind (that is: type a or type b, and in case of type b with the same real quadratic field). In case b there is just one geodesic ray associated with each cusp, in case a there is a continuum of geodesic rays associated with each cusp.

For $S\ell(3,C)$ and for the symplectic group of C. L. Siegel that acts on the space of complex 2 by 2 matrices Z with $\frac{Z - \bar{Z}}{2i}$ positive definite, somewhat similar results can be obtained.

For higher dimensions the number of cases to be distinguished seems to increase, although I do not definitely know that the number of possibilities that can actually be realized is that large.

Presumably one should in all cases where the rank is > 1 end up with some subgroups describing the cusp, which is such that this part of γ alone imposes rigidity on the full group. It also brings in a certain algebraic number field with which the full group is naturally associated.

10. It seems natural to formulate on the basis of what is known today, the following two conjectures:

A. An irreducible group Γ with $V(\Gamma\backslash S) < \infty$, can be deformed into a group whose matrix representation has only algebraic numbers from some algebraic number field as entries in the matrices, and which furthermore has bounded denominator.

B. In case the rank of S is > 1 and $\Gamma\backslash S$ noncompact, the group Γ is commensurable to some arithmetical group.

In A the deformation mentioned is of course trivial, unless S is the hyperbolic plane.

11. There are various interesting questions which we have not touched upon, which also bring out the distinction between rank one and rank > 1. Some of them have to do with the topological characteristics of the $\Gamma\backslash S$. Another question is whether the commutator subgroup $[\Gamma,\Gamma]$ of Γ is of finite index. Calabi-Vesentini [3], Matsushima [9] and Kazdan [7] have made very significant contributions.

Another question is whether it may be so that certain reducible spaces S do not have any irreducible groups with $V(\Gamma\backslash S) < \infty$. This would mean that the factors are in a sense incompatible. We cannot show that this is true for any S, but it is possible to show that certain reducible spaces S do not have any irreducible groups with $V(\Gamma\backslash S) < \infty$ and $\Gamma\backslash S$ noncompact. An example would be the reducible bounded symmetric domain $|z_1| < 1$, $|z_2|^2 + |z_3|^2 < 1$. The corresponding group is the product of $S\ell(2,R)$ (or $SO(2,1)$) and $SU(2,1)$. Another example is where G is the product of $S\ell(2,R)$ and $S\ell(3,R)$.

It is probable that the space S (or group G) influences the possible structure of groups Γ with $V(\Gamma\backslash S) < \infty$ profoundly. It is not known whether there are two different S that permit groups with the same structure. It is easy to see for example that $S\ell(2,R)$ and $S\ell(3,R)$ have no groups with finite volume of the fundamental domain, which are isomorphic.

References

1. A. Borel and Harish-Chandra, Arithmetic subgroups of algebraic groups, Ann. of Math. 75 (1967), pp. 485-535.

2. E. Calabi, On compact riemannian manifolds with constant curvature I, Proc. Symp. Pure Math. III (Differential Geometry) 1961, pp. 155-180.

3. ——— and E. Vesentini, On compact locally symmetric Kähler manifolds, Ann. of Math. 71 (1960), pp. 472-507.

4. R. Fricke and F. Klein, Vorlesungen über die Theorie der automorphen Funktionen. B. G. Teubner, Leipzig, 1912.

5. H. Garland and M. S. Raghunathan, Fundamental domains for lattices in rank one semisimple Lie groups, Yale University, Dept. of Math. preprint, 1968.

6. S. Helgason, Differential Geometry and Symmetric Spaces. Academic Press, New York, N. Y., 1962.

7. D. A. Kazdan, Connection of the dual space of a group with the structure of its closed subgroups, Funct. Analysis and its Appl. 1 (1967), pp. 63-65.

8. ——— and H. A. Margulies, Proof of the Selberg hypothesis, Mat. Sborn. 75 (1968), pp. 163-168.

9. Y. Matsushima, On the first Betti number of compact quotient spaces of higher dimensional symmetric spaces, Ann. of Math. 75 (1962), pp. 312-330.

10. G. D. Mostow, Quasiconformal mappings in n-space and rigidity of hyperbolic space forms, Inst. Hautes Études Sci. Publ. Math. 34 (1968), pp. 53-104.

11. I. I. Piatetsky-Shapiro, Discrete subgroups of analytic automorphisms of a polycylinder and automorphic forms, Dokl. Akad. Nauk. SSSR, 124 (1959), pp. 760-763.

12. A. Selberg, On discontinuous groups in higher dimensional symmetric spaces, Contributions to Function Theory, Bombay, 1960, pp. 147-164.

13. A. Weil, On discrete subgroups of Lie groups (I & II), Ann. of Math. 72 (1960), pp. 369-384 and 75 (1962), pp. 578-602.

The Corona Theorem

Lennart Carleson

1. Introduction.

Let C be the Banach algebra of analytic functions in $|z| < 1$ with a continuous extension to $|z| = 1$ under the uniform norm. The maximal ideal space can be identified with the disk $|z| \leq 1$. A particular consequence of this fact is that the ideal I generated by a finite number of functions $f_1(z), \ldots, f_n(z)$ is all of C if and only if

(1.1) $$|f_1(z)| + |f_2(z)| + \ldots + |f_n(z)| \geq \delta > 0$$

for $|z| < 1$ and a fixed $\delta > 0$.

It may be of interest to see how a proof of this fact, independent of the theory of maximal ideals, runs.

Let $L(f)$ be a linear functional vanishing on I. L can be identified with a measure μ on $|z| = 1$ and we have in particular the relations

$$\int_{-\pi}^{\pi} e^{i\nu\theta} f_j(e^{i\theta}) d\mu(\theta) = 0 , \quad \nu = 0, 1, 2, \ldots ; \quad j = 1, 2, \ldots, n .$$

By the F. and M. Riesz theorem it follows that there exist functions $F_j(z)$ in the Hardy class H^1 such that

$$F_j(e^{i\theta}) d\theta = f_j(e^{i\theta}) d\mu(\theta) , \quad F_j(0) = 0 .$$

It follows that $f_k(e^{i\theta}) F_j(e^{i\theta}) = f_j(e^{i\theta}) F_k(e^{i\theta})$ a.e. and so the meromorphic functions,

$$\frac{F_j(z)}{f_j(z)} = H(z)\ , \quad j = 1, 2, \ldots, n,$$

are identical. (1.1) implies that $H(z)$ is regular and belongs to H^1. This yields

$$\int d\mu(\theta) = \int H(e^{i\theta})d\theta = 2\pi \frac{F_j(0)}{f_j(0)} = 0$$

and so $1 \in I$ and $I = C$.

The corresponding result also holds for the Banach algebra H^∞ of bounded analytic functions. This result is known as the corona theorem (i.e. no corona exists!) and was proved in[1]. The proof there is quite complicated and it may be difficult to see the main lines. Quite recently, Hörmander [2] , [3] has simplified certain parts and introduced the new idea to use differential equations. Although the main difficulty remains unchanged (§ 4) the proof can now be given with greater clarity. Finally, the proof of [1] depends on a lemma on harmonic functions by Hall that can be built into the same general frame as other parts of the proof. We shall therefore here give a complete proof containing all necessay details.

2. Maximal theorems.

Let $f(x)$ be a non-negative real valued function defined on $(-\pi,\pi)$ and extend $f(x)$ with period 2π. We assume that $f(x) \in L^1(-\pi,\pi)$ and associate with $f(x)$ the following three functions:

$$f^*(x) = \sup_{t>0} \frac{1}{2t} \int_{x-t}^{x+t} f(u)du \ ;$$

$$f^*(x,y) = \sup_{t \geq y} \frac{1}{2t} \int_{x-t}^{x+t} f(u)du \ ;$$

$$u(x,r) = \frac{1}{2\pi} \int_{-\pi}^{\pi} \frac{1 - r^2}{1 + r^2 - 2r\cos(x-u)} f(u)du \ .$$

$f^*(x)$ is the Hardy-Littlewood maximal function and clearly $f^*(x) \geq f^*(x,y)$. A partial integration shows that **there is** an absolute constant A so that

(2.0) $\quad u(x,r) \leq A \; f^*(x,1-r)$.

We are interested in the sets where $f^*(x) > \lambda$ and $f^*(x,y) > \lambda$ in oneand two dimensions resp. and shall use the following elementary lemma.

Lemma. Let $S = \{I\}$ be a collection of intervals in $(-2\pi, 2\pi)$ covering a set E of measure $|E|$. Then there exists a countable set of disjoint intervals $I_j \in S$ such that

$$\sum_{j=1}^{\infty} |I_j| \geq \frac{1}{5} |E|$$

and such that every $I \subset$ some I_j^* where I_j^* is I_j increased in the scala 5:1 with the same centre as I_j .

Proof. Choose $I_1 \in S$ so that $|I_1| \geq \frac{1}{2} \sup |I|$. Next, choose I_2 not intersecting I_1 so that $|I_2| \geq \frac{1}{2}$ **sup** $|I|$ **under this condition.** I_3 is chosen similarly not intersecting $I_1 \cup I_2$ and so on.

Now, if $I \in S$, I has to intersect some I_k and we pick I_k with k minimal. This means $|I| \leq 2 \; |I_k|$ and since $I \cap I_k \neq \emptyset$, this implies $I \subset I_k^*$. Hence

$$|E| \leq \Sigma \; |I_k^*| \leq 5 \Sigma \; |I_k|,$$

as asserted.

Let us now study the set $E_\lambda = \{x \mid f^*(x) > \lambda\}$. If $x \in E_\lambda$ then there exists I containing x so that

(2.1) $\quad \int_I f(u)du > \lambda |I|$.

(2.1) defines a collection S of intervals. Choose I_j according to the lemma. It follows that

(2.2) $\quad \lambda |E_\lambda| \leq 5\lambda \; \Sigma |I_j| \leq 5 \; \Sigma \int_{I_j} f(u)du \leq 10 \int_{-\pi}^{\pi} f(u)du$.

We have thus obtained an estimate of $|E_\lambda|$.

Let us now in a similar way study the set F_λ of (x,y), $y \leq 1$, where $f^*(x,y) > \lambda$. We measure the size of this set with respect to a measure $\mu(x,y)$ and assume that μ satisfies the following condition:

For any given (x_0,y_0), let T_0 be triangle

(2.3) T_0: $y + |x-x_0|\ y_0$, $y > 0$.

We then assume

$$\mu(T_0) \leq A\ y_0 \ .$$

It is clear that $(x_0,y_0) \in F$ implies $T_0 \subset F$ and that if $I = (x_0-y_0,\ x_0+ y_0)$ satifies (2.1) then $(x_0,y_0) \in F$.

We again apply the lemma to the intervals I defined by (2.1). Every I is the base of a triangle T and to I_j corresponds T_j and to I_j^* T_j^*. Since every $I \subset$ some I_j^*, every $T \subset$ some T^*_j, i.e.

$$F_\lambda \subset \cup T_j^* \ .$$

This implies

(2.4) $\mu(F_\lambda) \leq \Sigma\mu(T_j^*) \leq A\ \Sigma|I_j^*| \leq 5\ A\ \Sigma\ |I_j| \leq \frac{1}{\lambda}\cdot 10A \int_{-\pi}^{\pi} f(u)du$.

This is the desired estimate of F_λ.

We now wish to estimate also the L^p-norms of our functions f^* and consider the mappings U:

$$U\colon f \to f^* \ .$$

Suppose that U quite generally is non-negative and sufficiently densly defined on one measure space $N(d\nu)$ with values f^* in another, $M(d\mu)$, and assume as is the case here

(2.5) $\|f^*\|_\infty = \|f\|_\infty$

and

(2.6) $\mu(\lambda) = \mu(\{m|f^*(m) > \lambda\}) \leq \frac{C}{\lambda}\cdot \int_N f(n)d\nu(n)$.

Finally, we assume $U(f + g) \leq UF + Ug$. It then foloows that

(2.7) $\int U(f)^p d\mu \leq \frac{p\cdot 2^p}{p-1}\cdot C\int f^p d\nu$.

This is a particular case of Marcinkiewicz' interpolation theorem.

To prove (2.7), take f and define $f_1 = f$ when $|f| \leq \lambda/2$. and $f_1 = 0$ otherwise. Let $f_2 = f - f_1$. Using the notations in (2.6) we have

$$\int |Uf|^p d\mu(\lambda) = \int_0^\infty \lambda^p d\mu(\lambda) = p \int_0^\infty \lambda^{p-1}\mu(\lambda)d\lambda \ .$$

Since $f_1^*(m) \leq \lambda/2$, the set where $f^*(m) \geq \lambda$ is included in the set where $f_2^*(m) \geq \frac{1}{2}\lambda$. Using obvious notations it follows

$$\int_0^\infty \lambda^{p-1}\mu(\lambda)d\lambda \leq \int_0^\infty \lambda^{p-1}\mu_2(\lambda/2)d\lambda \leq$$

$$\leq 2C \int_0^\infty \lambda^{p-2}d\lambda \int_{|f|>\frac{\lambda}{2}} f \cdot d\nu(n) = \frac{2C\cdot 2^{p-1}}{p-1} \int |f|^p d\nu \, .$$

If we specialize the above results and observe the inequality (2.0) we get the following result.

Theorem 1. Let $f(x) \in L^p$, $p > 1$, and let u be the harmonic function with boundary values f . Let μ be a positive measure such that

$$\mu(\{re^{i\theta}|, \ |\theta - \theta_0| \leq s, \ 1 - r \leq s\}) \leq As \ .$$

Then

$$\iint |u(re^{i\theta})|^p d\mu \leq \text{Const.} \int |f(x)|^p dx \ .$$

From considerations of maximal functions we can also deduce the following theorem on the minimum modulus of an analytic function.

Theorem 2. Let $f(z)$ be analytic in $|z| < 1$, $|f(z)| \leq 1$, and let $f(0) = a \neq 0$ be given. Let M be a large number and E the set of θ's for which $\inf_r |f(re^{i\theta})| \leq e^{-M}$. Then, given $\delta > 0$, there is a constant C only depending on a and δ so that for $M \geq C$

$$|E| \leq \delta \ .$$

Proof. We can factorize $f(z) = B(z) \cdot e^{-F(z)}$ where

$$B(z) = \prod_1^{\infty} \frac{b_\nu - z}{1 - z\bar{b}_\nu} \frac{\bar{b}_\nu}{|b_\nu|}$$

and $u(z) = \mathrm{Re}(F(z)) \geq 0$. Furthermore

$$\frac{1}{2\pi} \int u(re^{i\theta})d\theta \leq \log \frac{1}{|a|} .$$

From the inequality $u(re^{i\theta}) \leq A\, f^*(\theta)$ - assuming as we may that u has non-singular boundary values, since all estimates are uniform - it follows that the set E_1 belonging to $e^{-F(z)}$ and the constant $M/2$ satisfies $|E_1| \leq \mathrm{Const.}/M$. To get the result for $B(z)$ we write

$$(2.8) \qquad -\log|B(z)| = \Sigma \log \frac{1 - z\bar{b}_\nu}{z - b_\nu} = \int \log \frac{1 - z\bar{\zeta}}{z - \zeta}\, d\mu(\zeta) .$$

Let us first exclude the disks $|z - b_\nu| \leq c \cdot \delta(1 - |b_\nu|)$. They project on a set on $|z| = 1$ of measure

$$\leq c\delta\Sigma\, (1 - |b_\nu|) \leq c_1 \cdot \delta .$$

Outside these disks

$$-\log|B(z)| \leq C_\delta \int \frac{(1 - |z|^2)(1 - |\zeta|^2)}{|1 - z\bar{\zeta}|^2}\, d\mu(\zeta) .$$

Here the measure $(1 - |\zeta|^2)d\mu(\zeta)$ has uniformly bounded total mass and the integral increases if we project this mass on $|\zeta| = 1$. We get the mass μ^* and the inequality

$$-\log|B(z)| \leq C_\delta \int \frac{1 - |z|^2}{|e^{i\varphi} - z|^2}\, d\mu^*(\varphi) = u^*(z)$$

and on $u^*(z)$ we can again apply the maximal theorem. The theorem is proved.

It should be observed that the result does not hold if we replace $\log|f|$ by an arbitrary negative subharmonic function; it is essential that the logarithmic potential (2.8) is discrete with unit masses.

Let us also remark that we can take the minimum not only over radii but over a fixed sector with vertex at $e^{i\theta}$ inside the unit disk. This implies that the theorem can be used in similar conformally equivalent situations.

3. Let us now turn to the corona theorem. $f_1, f_2, \ldots, f_n \in H^\infty$ are given, $|f_j(z)| \leq 1$, and we wish to construct $p_1, p_2, \ldots, p_n \in H^\infty$ so that $\Sigma\, p_j f_j \equiv 1$. We shall prove that $|p_j(z)| \leq K(\delta)$ and are then allowed to assume that f_j are analytic in $|z| \leq 1$.

Let $\varphi_j(z) \in C^\infty$ in $|z| < 1$ and assume that $\varphi_j = 0$ when $|f_j| \leq \varepsilon$, where ε will be determined only depending on δ. We shall also have $0 \leq \varphi_j \leq 1$ and $\sum_1^n \varphi_j(z) \equiv 1$. We obviously have

$$1 \equiv \Sigma\, f_j(z)\, \frac{\varphi_j(z)}{f_j(z)} \,.$$

We shall construct p_j of the form

$$p_j(z) = \frac{\varphi_j(z)}{f_j(z)} + \sum_{k=1}^{n} a_{jk}(z)\, f_k(z) \,, \tag{3.1}$$

where a_{jk} satisfies $a_{jk} = -a_{kj}$ and $p_j(z)$ is analytic, i.e.

$$0 = \frac{\partial p_j}{\partial \bar z} = \frac{1}{f_j} \frac{\partial \varphi_j}{\partial \bar z} + \sum_k f_k \frac{\partial a_{jk}}{\partial \bar z} \,.$$

We choose a_{jk} so that

$$\frac{\partial a_{jk}}{\partial \bar z} = \left(\frac{\bar f_j}{f_k} \frac{\partial \varphi_k}{\partial \bar z} - \frac{\bar f_k}{f_j} \frac{\partial \varphi_j}{\partial \bar z} \right) \frac{1}{\sum_1^n |f_k|^2} \,. \tag{3.2}$$

$\partial p_j / \partial \bar z = 0$ follows since $\Sigma\, \varphi_j \equiv 1$. - Solving the corona problem is therefore equivalent to solving certain equations. Construct a so that

$$\frac{\partial a}{\partial \bar z} = F(z)$$

and $a(re^{i\theta}) \in L^1$ uniformly in r and $|\lim_{r\to 1} a(re^{i\theta})| \leq M$ a.e. where $F(z)$ has a bound Const. $\Sigma\,|\text{grad}\, \varphi_j|$.

<u>Theorem 3</u>. <u>The equation</u>

$$\frac{\delta a}{\delta \bar z} = F(z)$$

<u>where</u> $F(z) \in C^1$ <u>in</u> $|z| \leq 1$, has a solution $a(z)$ <u>in</u> $|z| \leq 1$, <u>such that</u> $a(re^{i\theta}) \in L^1$ <u>uniformly in</u> r <u>and</u> $|\lim a(re^{i\theta})| \leq M$ <u>a.e.</u> <u>if and only if</u>

(3.3) $$\left|\iint F(z)f(z)dxdy\right| \leq \frac{M}{2} \int_{|z|=1} |f(z)dx|$$

<u>for all polynomials</u> $f(z)$.

<u>Proof.</u> Assume first that $a(z)$ is such a solution. Then by a partial integration

$$\iint_{|z|<1} fFdxdy = \iint_{|z|<1} f(z)\frac{\delta a}{\delta \bar z}dxdy = \lim_{r\to 1}\frac{1}{2i}\int_{|z|=1} f(z)a(z)dz$$

which gives the desired inequality (3.3).

Conversely assume that (3.3) holds. It follows that there exists a function $\alpha(\zeta)$ such that $|\alpha(\zeta)| \leq M/2$ and

$$\iint F(z)f(z)dxdy = \int_{|\zeta|=1} \alpha(\zeta)\, f(\zeta)d\zeta .$$

We define, $|z| \neq 1$,

$$a(z) = \frac{1}{\pi i}\int_{|\zeta|=1} \frac{\alpha(\zeta)}{\zeta - z}\, d\zeta - \frac{1}{\pi i}\iint_{|\zeta|<1} \frac{F(\zeta)}{\zeta - z}\, d\xi\, d\eta .$$

It is well-known that $\delta a/\delta \bar z = F$ in $|z| < 1$ and $a(z)$ belongs to L^1 uniformly on circles $|z| = r$. Furthermore, if $|z| > 1$, $a(z) = 0$. Let z^* be the reflexion of z in the unit circle. It follows that if $|z_0| = 1$ and $z \to z_0$ along the radius

$$\lim_{z\to z_0} a(z) = \lim\,(a(z) - a(z^*)) = \lim \frac{1}{\pi i}\int \left(\frac{1}{\zeta - z} - \frac{1}{\zeta - z^*}\right)\alpha(\zeta)d\zeta = 2\alpha(z_0) \quad \text{a.e.}$$

Let us take the solutions a_{jk} of (3.2) obtained by the theorem and use them in (3.1). It follows that $p_j(z)$ is analytic in $|z|<1$ and belongs to L^1 uniformly on circles $|z| = r$. Finally, $\lim |p_j(re^{i\theta})| \leq \frac{1}{\epsilon} + nM$ and hence $p_j(z) \in H^\infty$.

To solve our problem we need now to be able to decide when an inequality (3.3) holds. We take the absolute value inside the integral sign and use the theorem on p.5. We find that U3.3) holds if (and with absolute values also only if)

(3.4) $$\iint_{Q_s} |F(z)|dxdy \leq A\cdot s$$

for all Q_s: $|\theta - \theta_0| \leq s$, $1 - r \leq s$.

Since $|F| \leq \text{Const.}\ \Sigma\ |\text{grad}\ \varphi_j|$, what now remains is to construct functions φ_j such that

(a) $\varphi_j(z) = 0$, when $|f_j(z) \leq \epsilon$,

(b) $0 \leq \varphi_j \leq 1$,

(c) $\Sigma\ \varphi_j = 1$,

(d) $|\text{grad}\ \varphi_j|$ satisfies (3.4) .

If we can satisfy instead of (c)

(c′) $\varphi_j = 1$ when $|f_j(z)| \geq \eta(\epsilon)$, $\eta(\epsilon)$ independent of f_j, $\eta(\epsilon) \to 0$, $\epsilon \to 0$, we only choose ϵ so that $\eta(\epsilon) = \frac{1}{n}\delta$ and then replace φ_j by $\varphi_j/\Sigma\ \varphi_j$. The denominator is never small.

We have in this way obtained a problem for a single analytic function $f = f_j$, and the construction of a corresponding φ satisfying (a), (b), (c′) and (d) will finish the proof.

4. We shall make the construction of the system $\varphi_j(z)$ in the case of the half-plane $y > 0$ - the modification to the case of the disk is quite obvious.

Let $\rho(z) = \frac{1}{y}$ and let $D(z_1, z_2)$ denote the distance between z_1, z_2 in $y > 0$ with respect to the density ρ . If $f(z)$ is analytic and $|f(z)| \leq 1$ in $y > 0$ then

(4.1) $$|f(z_1) - f(z_2)| \leq D(z_1, z_2)\ .$$

We consider $f(z)$ only in the square Q_0: $0 \leq z \leq 1$, $0 \leq y \leq 1$, and shall consider squares Q_n of generation n inside Q_0 of the form $h\ 2^{-n} \leq x \leq (h+1)2^{-n}$, $0 \leq y \leq 2^{-n}$, h a natural number. Let ϵ be our given number and choose the integer N so that $\epsilon > 2^{-N}$.

We first consider $f(z)$ in $Q_0 \setminus \bigcup_1^\infty Q_n = R_0$ and distinguish two cases.

I. There is a point $z_0 \in R_0$ so that $|f(z_0)| \geq 2^{-M}$ where M will be determined later. Let R_j be defined relatively Q_j as R_0 was defined relatively Q_0.

Consider the four squares $Q_2 \subset Q_0$ and in each R_2 we consider those dyadic squares with sides 2^{-N-2} which contain points where $|f(z)| \leq 2^{-N}$. From (4.1) it follows that $|f(z)| \leq \text{Const.}\ 2^{-N}$ in each such square. Those squares form a set A_2 which projects on a set A_2^* on $y = 0$. Next we consider the similar dyadic squares of side 2^{-N-3} in the R_3's whose projections fall outside A_2^*. They form A_3 with projection A_3^*. This construction goes on; for Q_ν the sides of the squares are $2^{-N-\nu}$.

In the set $A = \bigcup_2^\infty A_\nu$ we have $|f(z)| \leq \text{Const.}\ 2^{-N}$ and $A^* = \bigcup A^*_\nu$ is the projection of A. By Theorem 2 (modified to the case of a square), if M is not too large, but still so that $M \to \infty, N \to \infty$, the projection A^* has measure $|A^*| \leq \frac{1}{2}$ (say).

For each square S in A_ν we construct all dyadic squares of side $2^{-N-\nu}$ which project inside S^* (= the projection of S) and are situated in $y \geq 2^{-N-\nu}$. We now have the situation as follows

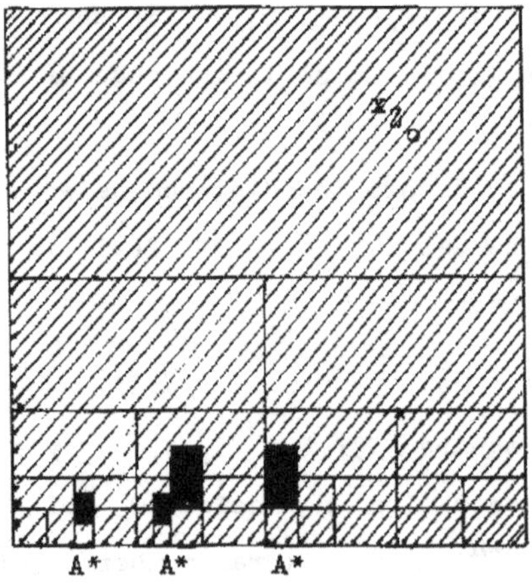

(N = 1)

Include also all dyadic squares of side 2^{-N-1} in Q_0 and Q_1 and the sides of Q_0. If we consider the length of the sides of all the squares constructed in this process it is clear that the length inside any square S, $x_0 \le x \le x_0 + \ell$, $0 < y < \ell$, is $\le$ Const. ℓ. It is also clear that their sides separate the set where $|f(z)| \ge 2^{-M}$ from the set where $|f(z)| \le 2^{-N}$, except perhaps in the remaining unshaded squares $Q^1, Q^2, \ldots$ in the figure, which project onto A*.

II. $|f(z)| < 2^{-M}$ in R_0. We then simply do no construction but proceed to consider the two squares Q_1 in Q_0.

It is clear that the above construction can be applied generally to any square Q by just a change of scale.

We start from Q_0. If I applies we do the construction defined and consider next the two squares Q_1. Next we consider the remaining squares (Q^i or Q_1) in a similar way. Every time that we use I, the total length of the remaining squares decreases by the factor $\frac{1}{2}$. It is therefore easy to see that the total length constructed inside any square of side ℓ is $\le$ Const. ℓ. Also, the system F separates the sets where $|f(z)| \le 2^{-N}$ and the set where $|f(z)| \ge 2^{-M}$.

Define a function $\psi(z)$ first as $= 0$ at all points where $|f(z)| \le 2^{-N}$ and also in every component of F containing such points. For the remaining points we define ψ as 2^{M+2} x the distance, with respect to $|dz|/y$, to the set F. It is clear by (4.1) that $\psi(z) \ge 1$ where $|f(z)| \ge 2^{-M+1}$. Define $\varphi(z) = \text{Min}\,(\psi(z), 1)$. It follows that

$$|\text{grad}\ \varphi| \le 2^{M+2}\frac{1}{y}$$

and that

$$\iint\limits_{\substack{x_0 \le x \le x_0+\ell \\ 0<y<\ell}} |\text{grad}\ \varphi|\,dxdy \le \text{Const.}\cdot\ell .$$

This completes the proof of our theorem.

References.

[1] Lennart Carleson: Interpolations by bounded analytic functions and the Corona problem. Ann. of math.76 (1962).

[2] Lars Hörmander: Generators for some rings of analytic functions. Bull.Amer.Math.Soc.73 (1967).

[3] Lars Hörmander: L^p-estimates for (pluri-) subharmonic functions. Math.Scand.20 (1967).

Polynomial Approximation

J. Wermer

§ 1. INTRODUCTION

Let X be a compact space and denote by $C(X)$ the Banach algebra of all continuous complex-valued functions on X. Fix $f_1,\dots,f_k$ in $C(X)$. We write

$$[f_1,\dots,f_k;X]$$

for the closed subalgebra generated by $f_1,\dots,f_k$ and the constants.

Our problem is to give conditions on the f_j and on X assuring that

$$[f_1,\dots,f_k;X] = C(X).$$

Evidently we want conditions that can be verified in special cases.

From the Stone-Weierstrass theorem we get at once:

<u>Proposition 1</u>: <u>Let</u> $g_1,\dots,g_n \in C(X)$ <u>and separate points on</u> X. <u>Then</u>

$$[g_1,\dots,g_n,\overline{g_1},\dots,\overline{g_n};X] = C(X).$$

Consider now the following example of our problem.

X is a compact subset of the space $\mathbb{C}^n$ of n complex variables. $z_1,\dots,z_n$ denote the coordinate functions. For what X does

$$[z_1,\dots,z_n;X] = C(X),$$

or in other words, on what sets X is every continuous function uniformly approximable by polynomials in the coordinates ?

Proposition 1 allows us to answer this question in a very special case. Denote by Σ_R the real subspace of $\mathbb{C}^n$, i.e. $\Sigma_R = \{(z_1,\dots,z_n) \mid z_j \text{ is real for each } j\}$.

Proposition 2: Let X be an arbitrary compact subset of Σ_R. Then

$$[z_1,\dots,z_n;X] = C(X).$$

To prove this, we note that $z_j = \bar{z}_j$ on Σ_R and hence that $[z_1,\dots,z_n;X] = [z_1,\dots,z_n,\bar{z}_1,\dots,\bar{z}_n;X]$.

For general compact $X \subset \mathbb{C}^n$ the Stone-Weierstrass theorem gives us no help, not even for $n = 1$, i.e. for subsets of the complex plane.

There is one immediate necessary condition which applies to all $X \subset \mathbb{C}^n$. A set X is called polynomially convex if for every $x^0 \in \mathbb{C}^n \setminus X$ we can find a polynomial Q with

$$|Q(x^0)| > \max_X |Q|.$$

Suppose the condition fails for some x^0. This means that:

$$(1.1) \qquad |P(x^0)| \leq \max_X |P|, \quad \text{all polynomials } P.$$

Thus the map $P \rightarrow P(x^0)$ from the algebra of polynomials into $\mathbb{C}$ is bounded, in the sup norm, and so admits representation by a measure:

$$P(x^0) = \int_X P d\mu,$$

μ a complex measure on X of finite total mass. Fix a polynomial R and fix j, $1 \leq j \leq n$, and put $P(z) = (z_j - x_j^0)R(z)$, where $x^0 = (x_1^0, \ldots, x_n^0)$. Then

$$0 = \int (z_j - x_j^0) R d\mu.$$

If now $[z_1, \ldots, z_n; X] = C(X)$, it follows that the measure

$$(z_j - x_j^0) d\mu$$

annihilates all continuous functions, and so is the zero measure. Hence the carrier of μ is contained in the set: $z_j - x_j^0 = 0$. Since this holds for each j, μ is carried by the point x^0, so $x^0 \in X$, contrary the assumption.

Thus the polynomial convexity of X is necessary for $[z_1,\dots,z_n;X] = C(X)$. Every compact subset of Σ_R is polynomially convex, as is easy to see directly.

Let us write $P(X)$ instead of $[z_1,\dots,z_n;X]$.

Consider now a compact plane set X, i.e. a subset of $\mathbb{C}^1$. For each x^o lying in a bounded complementary component of X we have

$$|P(x^o)| \leq \max_X |P|, \quad \text{all polynomials } P.$$

Hence, in order that X be polynomially convex we must have no bounded complementary components, which is the same as to say:

(A) The complement of X in $\mathbb{C}$ is connected.

(A) is thus necessary for $P(X) = C(X)$. It is not sufficient, as we see at once from the example: D is the closed unit disk. Here (A) holds, yet each function in $P(D)$ is analytic on the open disk and so $P(D) \neq C(D)$. To rule out such cases we ne d

(B) The interior of X is empty.

Indeed, (A) and (B) together are sufficient, (first shown by Lavrentieff):

<u>Proposition 3</u>: <u>Let</u> X <u>be a compact plane set satisfying (A) and (B). Then</u> $P(X) = C(X)$.

For $\mathbb{C}^n$, $n > 1$, we do not possess an analogue of Proposition 3, i.e. a necessary and sufficient condition on X for $P(X) = C(X)$.

In the following, we shall be concerned with generalisations of Propositions 1 and 2.

§ 2: A Perturbation Theorem

We begin with the following question: Let X be a compact set in the z-plane. We know that

$$[z,\bar{z};X] = C(X).$$

Is it still true that

$$[z,f;X] = C(X)$$

if f is near $\bar{z}$ in some sense ?

If "near" is taken to mean "uniformly close on X" the answer is No. To see this, let X be the closed unit disk and fix $\epsilon > 0$. Choose R in $C(X)$ with $|R| \leq \epsilon$ on X and $R = -\bar{z}$ in $|z| \leq \epsilon$. Then

$$[z,\bar{z}+R;X] \neq C(X),$$

yet $|(\bar{z}+R)-\bar{z}| \leq \epsilon$ everywhere on X.

If we interpret "near" as "close in Lipschitz norm", however, we can show:

<u>Proposition 4</u>: <u>Choose</u> R <u>in</u> $C(X)$ <u>with</u>

(2.1) $\quad |R(a)-R(b)| \leq k|a-b|$, <u>all</u> $a,b \in X$,

<u>where</u> k <u>is a constant</u> < 1. <u>Then</u>

$$[z,\bar{z}+R;X] = C(X).$$

<u>Proof</u>: Without loss of generality we may assume X is a disk. For else take a disk D containing X. By a result of Valentine, [1], R can be extended to a function $\tilde{R}$ on D with (2.1) holding for $\tilde{R}$ if $a,b \in D$. If the Proposition is true for D, then

$$[z,\bar{z}+\tilde{R};D] = C(D),$$

whence

$$[z,\bar{z}+R;X] = C(X),$$

i.e. we have the assertion for X.

Let us then take X to be a disk. Choose a complex measure μ on X with μ annihilating $\mathfrak{A} = [z,\bar{z}+R;X]$. We wish to show that $\mu = 0$. To this end fix $a \in \mathbb{C}$ such that

$$\int \frac{d|\mu|}{|z-a|} < \infty \quad . \tag{2.2}$$

First take $a \in X$. Put

$$\omega = (z-a)(\bar{z}+R(z)-\bar{a}-R(a)).$$

Then $\omega \in \mathfrak{A}$.

On the other hand,

$$\omega(z) = |z-a|^2+(z-a)(R(z)-R(a)),$$

and by (2.1) the second term is in modulus $<$ the first, except when $z = a$. Hence

$$\operatorname{Re}\omega(z) > 0 \quad \text{for} \quad z \neq a.$$

Let Π denote the closed half-plane $\operatorname{Re}\zeta \geq 0$. Put

$$P_n(\zeta) = \left[1 - \frac{1}{(1+\zeta)^n}\right] \cdot \frac{1}{\zeta}, \quad n = 1,2,\ldots$$

Then P_n is holomorphic on Π, hence uniformly approximable on compact subsets of Π by polynomials in ζ. It follows that $P_n(\omega)$ lies in $\mathfrak{A}$. Now

$$P_n(\zeta) \to \frac{1}{\zeta} \quad \text{as} \quad n \to \infty, \quad \text{for } \zeta \in \Pi, \ \zeta \neq 0, \quad \text{and}$$

$$|P_n(\zeta)| \leq \left[1 + \frac{1}{|1+\zeta|^n}\right]\frac{1}{|\zeta|} \leq \frac{2}{|\zeta|}, \quad \text{for } \zeta \in \Pi \text{ and all } n.$$

Put

$$f_n = (\bar{z}+R(z)-\bar{a}-R(a))P_n(\omega).$$

Then $f_n \in \mathfrak{A}$, and

$$f_n(z) \to \frac{1}{z-a}, \quad \text{all} \quad z \in D, \ z \neq a, \quad \text{and}$$

$$|f_n(z)| \leqq \frac{2}{|z-a|}, \quad \text{all} \quad z \in D.$$

By dominated convergence it follows, using (2.2), that

$$\int f_n d\mu \to \int \frac{d\mu}{z-a}.$$

But $f_n \in \mathfrak{A}$ and μ annihilates $\mathfrak{A}$. Hence

(2.3) $$\int \frac{d\mu}{z-a} = 0.$$

Next, take $a \notin X$. Then $\frac{1}{z-a} \in [z;X]$, hence $\in \mathfrak{A}$ and so (2.3) holds there too.

It is well known, and follows directly by Fubini's theorem, that (2.2) holds for almost all a. Hence both (2.2) and (2.3) hold for almost all a.

Choose now a smooth function g defined and of compact support on $\mathbf{C}$. Stokes' Theorem gives

(2.4) $$g(\zeta) = -\frac{1}{\pi}\iint \frac{\partial g}{\partial \bar{z}}(z)\frac{dxdy}{z-\zeta}, \qquad \zeta \in \mathbf{C}.$$

Hence

$$\int g(\zeta)d\mu(\zeta) = -\frac{1}{\pi}\iint \frac{\partial g}{\partial \bar{z}}(z)\left\{\int \frac{d\mu(\zeta)}{z-\zeta}\right\} dxdy$$

The integrand vanishes for a.a. z by (2.3). Thus μ annihilates g. Since such g are dense in $C(X)$, $\mu = 0$, as desired. It follows that $\mathfrak{A} = C(X)$, and we are done.

<u>Note</u>: For X a disk, Proposition 4 is proved in [2]. See also [3] for related work.

§ 3. Complex Tangents

Let now Σ be any smooth real submanifold of an open subset of $\mathbb{C}^n$ and $x \in \Sigma$. The tangent space T_x to Σ at x can be regarded as a real-linear subspace of $\mathbb{C}^n$. By a complex line we mean the set of all multiples by complex scalars of a fixed non-zero vector in $\mathbb{C}^n$. Thus a complex line is a 2-dimensional real-linear subspace L of $\mathbb{C}^n$ invariant under multiplication by i.

Definition: A complex tangent to Σ at x is a complex line contained in T_x.

Examples. a) Let Σ be a complex analytic manifold in $\mathbb{C}^n$. Then the tangent space to Σ at a given point x is invariant under multiplication by i. Hence Σ has a complex tangent at each point.

b) Σ_R coincides with its tangent space at each point, so Σ_R has no complex tangents.

c) Σ is the image in $\mathbb{C}^2$ of the disk: $|z| < 1$ under the map : $z \to (z, f(z))$, where f is a smooth complex-valued function.

Σ has a complex tangent at the point $(a, f(a))$ if and only if $\frac{\partial f}{\partial \bar{z}}(a) = 0$.

d) Let $f_1, \dots, f_n$ be smooth complex-valued functions defined in a neighborhood U of 0 in $\mathbb{R}^k$ such that $f = (f_1, \ldots, f_n)$ is a regular map of U into $\mathbb{C}^n$. Let $\Sigma = f(U)$.

Σ has no complex tangent at $f(0)$ if and only if the matrix

$$\begin{pmatrix} \frac{\partial f_1}{\partial x_1}, \ldots, \frac{\partial f_1}{\partial x_k} \\ \vdots \\ \frac{\partial f_n}{\partial x_1}, \ldots, \frac{\partial f_n}{\partial x_k} \end{pmatrix}$$

has rank k at 0. Call this condition (1).

<u>Note</u>: Assume now that the f_j are real-analytic and that (1) holds. There exists a complex-analytic extension f^* of the map f to a neighborhood U^* of 0 in $\mathbb{C}^k$. Because of (*) it is no loss of generality to assume that f^* maps U^* biholomorphically on a piece of complex submanifold Σ^* of $\mathbb{C}^n$. Σ^* has complex dimension k, and Σ is the subset of Σ^* corresponding under f^* to $\mathbb{R}^k$.

Thus, in the real analytic case, a manifold Σ without complex tangents is locally equivalent to $\mathbb{R}^k$ under a biholomorphic map.

e) Let Σ be a k-dimensional smooth real submanifold of $\mathbb{C}^n$ with $k > n$. (Recall that $\mathbb{C}^n = \mathbb{R}^{2n}$). Then Σ has a complex tangent at each point. The reason is that a k-dimensional real-linear subspace of $\mathbb{C}^n$ must contain a complex line if $k > n$.

The notion of complex tangent allows us to give a sufficient condition on a compact set X in $\mathbb{C}^n$ in order that the algebra of polynomials be dense in $C(X)$.

Theorem 1: Let Σ be a smooth submanifold of some open subset of $\mathbb{C}^n$ such that Σ has no complex tangents, and let K be a polynomially convex compact subset of Σ. Then $P(K) = C(K)$.

Proposition 2 is evidently a special case of this result.

Put now $|\zeta|$ = Euclidean norm of ζ, for $\zeta \in \mathbb{C}^n$.

Theorem 2: Let X be an arbitrary compact subset of $\mathbb{C}^n$. Let $R = (R_1,\ldots,R_n)$ be a smooth vector-valued function defined in a neighborhood $\mathcal{N}$ of X in $\mathbb{C}^n$. Assume there is a constant $k < 1$ such that

(2) $\quad |R(z)-R(z')| \leq k|z-z'|$ for all $z,z' \in \mathcal{N}$.

Then

$$\left[z_1,\ldots,z_n,\bar{z}_1+R_1,\ldots,\bar{z}_n+R_n;X\right] = C(X).$$

This result gives a perturbation of Proposition 1, and reduces to Proposition 1 when $R = 0$. One would like to show that the conclusion of Theorem 2 holds with (2) required only for $z,z' \in X$ and with no smoothness assumptions on R. When $n = 1$, Proposition 4 shows this.

For 2-dimensional Σ results of the type of Theorem 1 are given in [4], [5]. For k-dimensional Σ, $k > 2$, the method of proof is due to Hörmander and results closely related to Theorem 1 are given in [6], [7] by R. Nirenberg and R.O. Wells.

Theorem 2 and a generalisation of Theorem 1 to the case when there exist complex tangents to Σ are given by Hörmander and the author in [8].

We proceed to prove Theorem 1. In the proof of Lemma 3 we follow Nirenberg and Wells [7], as the proof given there is simpler than the proof in [8].

Finally, we shall sketch the way that leads from Theorem 1 to Theorem 2.

Note: The reader can find the notions from function theory in $\mathbb{C}^n$ which are used in the proof of Theorem 1 discussed, e.g., in [9], Chapter II.

Note: The precise sense of "smooth" in the hypotheses of the two theorems will appear in the proofs.

§ 4. Proof of Theorem 1

To prove the theorem, it suffices to show that every smooth function defined on Σ, when restricted to K, lies in P(K).

We choose such a smooth function u on Σ and hold it fixed from now on. We shall appeal to the following:

Oka-Weil Theorem: ([9], Theorem 2.7.7).
If K is a polynomially convex compact set in $\mathbb{C}^n$ and if h is holomorphic in a neighborhood of K in $\mathbb{C}^n$, then $h|K$ lies in P(K).

We thus need only approximate u by functions holomorphic in some neighborhood of K.

Notations: Ω a region in $\mathbb{C}^n$, f a smooth function defined in Ω. Put

$$\bar{\partial} f = \sum_{j=1}^{n} \frac{\partial f}{\partial \bar{z}_j} d\bar{z}_j$$

Thus f is holomorphic in Ω if and only if $\bar{\partial} f = 0$ in Ω.

Let F be a smooth 1-form in Ω, $F = \sum_{j=1}^{n} F_j d\bar{z}_j$. Put

$$\bar{\partial} F = \sum_{j=1}^{n} \bar{\partial} F_j \wedge d\bar{z}_j \ .$$

Note that $\bar\partial^2 = 0$, i.e. $\bar\partial(\bar\partial f) = 0$ for all f.

To get a holomorphic approximation to u, we shall do the following:

Step 1: To construct for each $\epsilon > 0$ a certain domain ω_ϵ in $\mathbb{C}^n$ with $K \subset \omega_\epsilon$ such that $\omega_\epsilon \to K$ as $\epsilon \to 0$.

Step 2: To find an extension U_ϵ of u to ω_ϵ such that $\bar\partial U_\epsilon$ is "small" in ω_ϵ.

Step 3: To find a function V_ϵ in ω_ϵ such that $\bar\partial V_\epsilon = \bar\partial U_\epsilon$ in ω_ϵ and $\sup_K |V_\epsilon| \to 0$ as $\epsilon \to 0$.

Once step 3 is done, we can write

$$U_\epsilon = (U_\epsilon - V_\epsilon) + V_\epsilon \quad \text{in } \omega_\epsilon.$$

Since $\bar\partial(U_\epsilon - V_\epsilon) = 0$ in ω_ϵ, $U_\epsilon - V_\epsilon$ is holomorphic in ω_ϵ and since $\sup_K |V_\epsilon| \to 0$, this holomorphic function approximates $u = U_\epsilon$ as closely as we please on K.

Lemma 1: Let Σ be a real submanifold of an open subset of $\mathbb{C}^n$ of class C^2. Let $d(x,\Sigma)$ be the distance from x to Σ. If Σ has no complex tangents, then there is a neighborhood ω of Σ such that d^2 is smooth and strongly pluri-subharmonic in ω, i.e.

$$L(z;\xi) = \sum_{i,j=1}^{n} \frac{\partial^2(d^2(z,\Sigma))}{\partial z_i \, \partial \bar z_j} \xi_i \bar\xi_j > 0$$

at all $z \in \omega$, $\xi = (\xi_1,\ldots,\xi_n) \neq 0$.

<u>Proof</u>: Fix $\zeta \in \Sigma$. We assert $L(\zeta;\xi) > 0$ for $\xi \neq 0$. With no loss of generality we can take $\zeta = 0$.

Let T be the tangent space to Σ at 0. Then

(3) $$d^2(z,\Sigma) = d^2(z,T)+o(|z|^2),$$

as one may verify. Also

$$d^2(z,T) = H(z)+\mathrm{Re}A(z)$$

where $H(z) = \sum_{j,k=1}^{n} h_{jk} z_j \bar{z}_k$ is hermitean-symmetric and A is a quadratic polynomial in z. Then

$$d^2(w,T)+d^2(iw,T) = 2H(w).$$

Now if $w \neq 0$, either $w \notin T$ or $iw \notin T$, since by hypothesis T contains no complex line. So if $w \neq 0$, $H(w) > 0$. Now

$$L(0;\xi) = \sum_{i,j=1}^{n} \left.\frac{\partial^2(d^2(z,\Sigma))}{\partial z_i \partial \bar{z}_j}\right|_{z=0} \cdot \xi_i \bar{\xi}_j = H(\xi).$$

Thus $L(\zeta;\xi) > 0$ when $\xi \neq 0$.

It follows by continuity that $L(z;\xi) > 0$ for z in some neighborhood of Σ and $\xi \neq 0$.

q.e.d.

Let Σ be as in the preceding Lemma and let K be a compact subset of Σ. By Lemma 1, K has a neighborhood ω in $\mathbb{C}^n$ where $d^2 = d^2(z,\Sigma)$ is strongly pluri-subharmonic.

Choose a C^∞ function φ in $\mathbb{C}^n$ having compact support in ω with $\varphi = 1$ in a neighborhood of K and $0 \leq \varphi \leq 1$ everywhere. For $0 < \varepsilon$, put

$$\omega_\varepsilon = \{z \in \mathbb{C}^n \mid d^2 - \varepsilon^2 \varphi < 0.\}$$

Lemma 2: For all small $\epsilon > 0$, ω_ϵ is a pseudo-convex open set with $K \subset \omega_\epsilon$.

Proof: $d^2 - \epsilon^2\varphi = -\epsilon^2$ on K, so $K \subset \omega_\epsilon$, and evidently ω_ϵ is an open set.

Fix ϵ_0. If $z \in \omega_{\epsilon_0}$, $d^2(z) < \epsilon_0^2\varphi(z)$, whence $z \in \operatorname{supp}\varphi$. So $\omega_{\epsilon_0} \subset\subset \omega$.

The Levi form of d^2 has a positive lower bound on ω_{ϵ_0} for $|\zeta| = 1$.

Hence ω_ϵ is pseudo-convex for small ϵ.

Q.E.D.

This completes Step 1.

Lemma 3: Let Σ be a real submanifold of some open subset of $\mathbb{C}^n$ of class C^e. Assume Σ has no complex tangents. Fix a compact set $K \subset \Sigma$.

Let u be a function of class C^e defined on Σ. Then $\exists$ a function U of class C^1 in $\mathbb{C}^n$ with:

(i) $U \equiv u$ on K.

(ii) $\exists$ constant C with

$$\left| \frac{\partial U}{\partial \bar{z}_j}(z) \right| \leq C \cdot d(z,\Sigma)^{e-1}, \quad \text{all } z, \; j = 1,\ldots,n$$

<u>Proof</u>: (as in [7]). We first perform the extension locally.

Fix $x_0 \in \Sigma$. Choose an open set Ω in $\mathbb{C}^n$ such that $x_0 \in \Omega$ and

$$\Sigma \cap \Omega = \left\{ x \in \Omega \mid \rho_1(x) = \ldots = \rho_m(x) = 0 \right\},$$

where each ρ_j is of class C^e in Ω and such that u has an extension to $C^e(\Omega)$, again denoted u.

We assert $\exists$ a neighborhood ω_0 of x_0 and $\exists$ integers $\nu_1, \nu_2, \ldots, \nu_n$ such that the vectors

$$\left(\frac{\partial \rho_{\nu_j}}{\partial \bar{z}_1}, \ldots \frac{\partial \rho_{\nu_j}}{\partial \bar{z}_n} \right)_x, \quad j = 1,\ldots,n$$

form a basis for $\mathbb{C}^n$ for each $x \in \omega_0$.

Put $\xi_\nu = \left(\frac{\partial \rho_\nu}{\partial \bar{z}_1}, \ldots, \frac{\partial \rho_\nu}{\partial \bar{z}_k} \right)_{x_0}$, $\nu = 1,\ldots,m$.

Suppose that $\xi_1, \ldots, \xi_m$ fail to span $\mathbb{C}^n$. Then $\exists$ $c = (c_1,\ldots,c_n) \neq 0$ with $(c, \xi_\nu) = 0$, all ν. Thus $\sum_{j=1}^{n} c_j \frac{\partial \rho_\nu}{\partial \bar{z}_j} = 0$, $\nu = 1,\ldots,m$. In other words the tangent vector to $\mathbb{C}^n$ at x_0:

$$\sum_{j=1}^{n} c_j \frac{\partial}{\partial \bar{z}_j}$$

annihilates $\varrho_1,\ldots,\varrho_o$ and hence is tangent to Σ. Thus Σ has a complex tangent at x_o, which is contrary to assumption.

Hence $\xi_1,\ldots,\xi_m$ spare $\mathbb{C}^n$, and so we can find $\nu_1,\ldots,\nu_n$ with $\xi_{\nu_1},\ldots,\xi_{\nu_n}$ linearly independent. By continuity, then the vectors

$$\left(\frac{\partial \varrho_{\nu_j}}{\partial \bar{z}_1},\ldots,\frac{\partial \varrho_{\nu_j}}{\bar{z}_n}\right)_x \quad , \quad j = 1,\ldots,n$$

are linearly independent, and so form a basis for $\mathbb{C}^n$, for all x in some neighborhood of x_o. This was the assertion.

Relabel $\varrho_{\nu_1},\ldots,\varrho_{\nu_n}$ to read: $\varrho_1,\ldots,\varrho_n$. Define functions $h_1,\ldots,h_n$ in ω_o by:

$$\left(\frac{\partial u}{\partial \bar{z}_1},\ldots,\frac{\partial u}{\partial \bar{z}_n}\right)(x) = \sum_{i=1}^{n} h_i(x)\left(\frac{\partial \varrho_i}{\partial \bar{z}_1},\ldots,\frac{\partial \varrho_i}{\partial \bar{z}_n}\right)_x, x \in \omega_o .$$

Solve for $h_i(x)$. All the coefficients in this $n \times n$ system of equations are of class $e-1$, so $h_i \in C^{e-1}(\omega_o)$. We have

$$\bar{\partial} u = \sum_{i=1}^{n} h_i \bar{\partial} \varrho_i \quad \text{in } \omega_o .$$

Put

$$u_1 = u - \sum_{i=1}^{n} h_i \rho_i . \quad \text{So} \quad u_1 = u \quad \text{on} \quad \Sigma, \quad \text{and}$$

$$\bar{\partial} u_1 = \bar{\partial} u - \sum_{i=1}^{n} h_i \bar{\partial}\rho_i - \sum_{i=1}^{n} \bar{\partial} h_i \cdot \rho_i = - \sum_{i=1}^{n} \bar{\partial} h_i \cdot \rho_i$$

In the same way in which we got the h_i we can find functions h_{ij} in $C^{e-2}(\omega_o)$ with

$$\bar{\partial} h_i = \sum_{j=1}^{n} h_{ij} \bar{\partial}\rho_j , \quad i = 1,\dots,n$$

Since $\bar{\partial}\rho_1,\dots,\bar{\partial}\rho_n$ are linearly independent at each point of ω_o, the same is true of the (0,2)-forms $\bar{\partial}\rho_j \wedge \bar{\partial}\rho_i$ with $i < j$.

$$0 = \bar{\partial}^2 u = \bar{\partial}\Big(\sum_{i=1}^{n} h_i \bar{\partial}\rho_i\Big) = \sum_i \Big(\sum_j h_{ij} \bar{\partial}\rho_j\Big) \wedge \bar{\partial}\rho_i$$

$$= \sum_{i<j} (h_{ij} - h_{ji}) \cdot \bar{\partial}\rho_j \wedge \bar{\partial}\rho_i$$

Hence $h_{ij} = h_{ji}$ for $i < j$. Put

$$u_2 = u_1 + \frac{1}{2!} \sum_{i,j} h_{ij} \rho_i \rho_j . \quad \text{So} \quad u_2 = u \quad \text{on} \quad \Sigma \quad \text{and}$$

$$\bar{\partial} u_2 = - \sum_i \bar{\partial} h_i \cdot \rho_i + \frac{1}{2!} \sum_{i,j} \bar{\partial}(h_{ij}) \rho_i \rho_j + R, \quad \text{where}$$

$$R = \frac{1}{2} \sum_{i,j} h_{ij} \rho_i \bar{\partial}\rho_j + \frac{1}{2} \sum_{i,j} h_{ij} \rho_j \bar{\partial}\rho_i$$

$$= \frac{1}{2} \sum_i \bar{\partial} h_i \cdot \rho_i + \frac{1}{2} \sum_j \bar{\partial} h_j \rho_j , \quad \text{so}$$

$$\bar{\partial} u_2 = \frac{1}{2!} \sum_{i,j} \bar{\partial} h_{ij} \cdot \rho_i \rho_j .$$

We define inductively functions h_I on ω_0, I a multi index, by

$$\bar{\partial} h_I = \sum_{j=1}^{n} h_{Ij} \bar{\partial} \rho_j$$

and we define functions u_N, $N = 1,2,\ldots,e-1$, by

$$u_N = u_{N-1} + \frac{(-1)^N}{N!} \sum_{|I|=N} h_I \rho_I ,$$

where $I = (\beta_1,\ldots,\beta_n)$, $|I| = \sum \beta_i$, $\rho_I = \rho_1^{\beta_1} \cdots \rho_n^{\beta_n}$.
Then $h_I \in C^{e-N}(\omega_0)$ if $|I| = N$, and $u_N \in C^{e-N}(\omega_0)$.

We verify:

$$\bar{\partial} u_N = \frac{(-1)^N}{N!} \sum_{|I|=N} \bar{\partial} h_I \cdot \rho_I , \quad \text{each } N.$$

Also there is a constant C such that $|\rho_I(z)| \leq C d(z,\Sigma)^N$ in ω_0 if $|I| = N$, and hence there is a constant C_1 with

$$\left| \frac{\partial u_N}{\partial \bar{z}_j}(z) \right| \leq C_1 d(z,\Sigma)^N, \quad j = 1,\ldots,n, \quad z \in \omega_0 .$$

In particular, $u_{e-1} \in C^1(\omega_0)$, $u_{e-1} = u$ on Σ, and

$$\left| \frac{\partial u_{e-1}}{\partial \bar{z}_j} \right| \leq C_1 d(z,\Sigma)^{e-1}, \quad C_1 \text{ depending on } \omega_0 .$$

Also, $u = 0$ on an open subset of ω_0 implies $u_{e-1} = 0$ there.

For each $x_0 \in K$ we now choose a neighborhood ω_{x_0} in $\mathbb{C}^n$ of the above type. Finitely many of these neighborhoods, say $\omega_1, \dots, \omega_g$, cover K.

Choose $\chi_1, \dots, \chi_g \in C^\infty(\mathbb{C}^n)$ with $\operatorname{supp} \chi_\alpha \subset \omega_\alpha$, $0 \leq \chi_\alpha \leq 1$ and $\sum_{\alpha=1}^{g} \chi_\alpha = 1$ on K.

By the above construction, chose U_α in $C^1(\omega_\alpha)$ with $U_\alpha = \chi_\alpha u$ in $\Sigma \cap \omega_\alpha$, $\operatorname{supp} U_\alpha \subset \operatorname{supp} \chi_\alpha u$ and

(4) $$\left| \frac{\partial U_\alpha}{\partial \bar{z}_j}(z) \right| \leq C_\alpha \cdot d(z,\Sigma)^{e-1}, \quad z \in \omega_\alpha, \; j = 1,\dots,n.$$

Since $\operatorname{supp} U_\alpha \subset \omega_\alpha$ we can define $U_\alpha = 0$ outside ω_α to get a C^1-function in the whole space, and (4) holds for all z in $\mathbb{C}^n$.

Put $U = \sum_{\alpha=1}^{g} U_\alpha$.

Then $U \in C^1(\mathbb{C}^n)$ and for $z \in K$

$$U(z) = \sum_{\alpha=1}^{g} U_\alpha(z) = \sum_{\alpha=1}^{g} \chi_\alpha(z) u(z) = u(z) \sum_{\alpha} \chi_\alpha = u(z).$$

For every z, $\frac{\partial U}{\partial \bar{z}_j}(z) = \sum_{\alpha=1}^{g} \frac{\partial U_\alpha}{\partial \bar{z}_j}(z)$, so by (4),

$$\left| \frac{\partial U}{\partial \bar{z}_j}(z) \right| \leq g \cdot C \cdot d(z,\Sigma)^{e-1}, \quad \text{where } C = \max_{1 \leq \alpha \leq g} C_\alpha.$$

q.e.d.

This completes Step 2.

$\bar{\partial}$-Lemma: Let Ω be a pseudo-convex open set with diameter $\delta < \infty$ in $\mathbb{C}^n$. Fix a form f of type $(0,1)$ in $L^2_{0,1}(\Omega)$, (i.e. $f = \sum_{j=1}^{n} f_j d\bar{z}_j$, $|f|^2 = \sum_{j=1}^{n} |f_j|^2$ and $\int_\Omega |f|^2 dV = \|f\|^2 < \infty$). Assume $\bar{\partial} f = 0$.

Then there is $w \in L^2(\Omega)$ with $\bar{\partial} u = f$ and

(5) $$\|w\|^2_{L^2(\Omega)} \leq e \cdot \delta^2 \cdot \|f\|^2_{L^2(\Omega)}$$

Note: We take $\bar{\partial} u$ in the sense of distribution theory.

Proof: See L. Hörmander, [10], Theorem 2.2.3. This result sharpens earlier results of Morrey [11] and Kohn [12].

Lemma 4: Let K, Σ, u, U be as in Lemma 3. Then for all small $\varepsilon > 0$ there exists w_ε in $L^2(\omega_\varepsilon)$, (ω_ε as in Lemma 2) with

(6) $$\bar{\partial} w_\varepsilon = \bar{\partial} U \text{ in } \omega_\varepsilon \text{ and}$$

(7) $$\|w_\varepsilon\|_{L^2(\omega_\varepsilon)} = O(\varepsilon^{e-1+n-\frac{1}{2}k}),\ k = \dim \Sigma.$$

Proof: By (ii) of Lemma 3,

(8) $$\left|\frac{\partial U}{\partial \bar{z}_j}(z)\right| \leq C d(z,\Sigma)^{e-1}, \quad \text{all } z \in \omega_\varepsilon, \text{ all } j.$$

Now $z \in \omega_\varepsilon \Rightarrow d^2(z,\Sigma) < \varepsilon^2 \varphi(z) \Rightarrow d(z,\Sigma) < \varepsilon$. Thus ω_ε lies in the ε-tube round Σ. Fix ε_0. Then $\omega_\varepsilon \subset \overline{\omega}_{\varepsilon_0}$,

a compact subset of $\mathbb{C}^n$ independent of ε, if $\varepsilon < \varepsilon_0$.
Hence the volume of $\omega_\varepsilon = O(\varepsilon^{2n-k})$, where $k = \dim \Sigma$.
(8) gives that

$$(9) \qquad \left|\frac{\partial U}{\partial \bar{z}_j}\right| < C\,\varepsilon^{e-1} \quad \text{in } \omega_\varepsilon. \quad \text{Hence}$$

$$\|\bar{\partial} U\|^2_{L^2(\omega_\varepsilon)} = \int_{\omega_\varepsilon} \left(\sum_{j=1}^{n} \left|\frac{\partial U}{\partial \bar{z}_j}\right|^2\right) dV \leq nC^2 \varepsilon^{2e-2} \operatorname{vol}(\omega_\varepsilon)$$

$= O(\varepsilon^{2e-2+2n-k})$. Also $\bar{\partial}(\bar{\partial} U) = 0$.

By the $\bar{\partial}$-Lemma, $\exists\, w_\varepsilon \in L^2(\omega_\varepsilon)$ with $\bar{\partial} w_\varepsilon = \bar{\partial} U$
and $\|w_\varepsilon\|^2_{L^2(\omega_\varepsilon)} < e(\operatorname{diam} \omega_{\varepsilon_0})^2 \|\bar{\partial} U\|^2_{L^2(\omega_\varepsilon)}$. So

$$\|w_\varepsilon\|_{L^2(\omega_\varepsilon)} = O(\varepsilon^{e-1+n-\frac{k}{2}}).$$

q.e.d.

<u>Lemma 5</u>: Let $B = \{ z \in \mathbb{C}^n \,|\, |z| < 1\}$ and $u \in C^\infty$. Then

$$|u(0)| \leq C\left\{\|u\|_{L^2(B)} + \sup_B \left(\max_{1\leq j\leq n} \left|\frac{\partial u}{\partial \bar{z}_j}\right|\right)\right\}.$$

<u>Proof</u>: Put $E(x) = \dfrac{C}{|x|^{2n-2}}$ = fundamental solution of the Laplace operator in $\mathbb{R}^{2n}$.

Choose $\chi \in C_0^\infty(B)$ with $\chi = 1$ in $|z| < \frac{1}{2}$. Then

$$u(0) = u\chi(0) = \int \Delta\,(\chi u) \cdot E(x)\,dx$$

$$= \int \Delta\chi \cdot u(x) E(x)\,dx + \int \Delta u \cdot \chi(x) E(x)\,dx$$

$$+2\int(\operatorname{grad}\chi, \operatorname{grad} u)E(x)dx = I_1+2I_2+I_3.$$

$$\int \chi_{x_i} u_{x_i} E dx = \int u_{x_i}(\chi_{x_i}E)dx = -\int u(\chi_{x_i}E)_{x_i}dx.$$

So $I_2 = -\int u(x)\sum_{i=1}^{n}(\chi_{x_i}E)_{x_i}dx$. Also

$$I_3 = 4\int\Big(\sum_j \frac{\partial^2 u}{\partial z_j \partial \bar{z}_j}\Big)\chi(x)E(x)dx = 4\sum_j\int\frac{\partial}{\partial z_j}\Big(\frac{\partial u}{\partial \bar{z}_j}\Big)\chi E dx$$

$$= -4\sum_j\int\frac{\partial u}{\partial \bar{z}_j}\frac{\partial}{\partial z_j}(\chi E)dx = -4\int\sum_j\frac{\partial u}{\partial \bar{z}_j}\cdot\frac{\partial}{\partial z_j}(\chi E)dx.$$

Since $\chi = 1$ in $|z| < \frac{1}{2}$, $\Delta\chi = 0$ and $(\chi_{x_i}E)_{x_i} = 0$ in $|z| < \frac{1}{2}$, and so

$|I_1| \leq C\|u\|_{L^2(B)}$, $|I_2| \leq C\|u\|_{L^2(B)}$. Also

$$|I_3| \leq C \sup_B\Big(\max_{1\leq j\leq n}\Big|\frac{\partial u}{\partial \bar{z}_j}\Big|\Big).$$

Hence the assertion.

<u>Corollary</u>: Fix $z_0 \in \mathbb{C}^n$. Put $B_\varepsilon = \{z \in \mathbb{C}^n \mid |z-z_0| < \varepsilon\}$, and $w \in L^2(B_\varepsilon)$. Suppose $\bar\partial w$ is continuous. Then w is continuous and

$$(10)\quad |w(z_0)| \leq C\Big\{\varepsilon^{-n}\cdot\|w\|_{L^2(B_\varepsilon)} + \varepsilon\cdot\sup_{B_\varepsilon}\Big(\max_{1\leq j\leq n}\Big|\frac{\partial w}{\partial \bar{z}_j}\Big|\Big)\Big\}.$$

<u>Proof</u>: We do not prove the continuity of w. Without loss of generality, $z_0 = 0$. Put $u(z) = w(\varepsilon z)$ for $z \in B$.

With dV the element of volume:

$$\|u\|^2_{L^2(B)} = \iint_B |w(\varepsilon z)|^2 dV = \iint_{B_\varepsilon} |w(\zeta)|^2 \varepsilon^{-2n} dV, \quad \text{so}$$

$$\|u\|_{L^2(B)} = \varepsilon^{-n} \|w\|_{L^2(B_\varepsilon)}.$$

Also $\frac{\partial u}{\partial \bar{z}_j}(z) = \varepsilon \frac{\partial w}{\partial \bar{z}_j}(\varepsilon z)$, $z \in B$. By Lemma 5, then,

$$|w(0)| = |u(0)| \le C\left\{\varepsilon^{-n}\|w\|_{L^2(B_\varepsilon)} + \varepsilon \sup_{B_\varepsilon}\left\{\max_{1\le j\le n}\left|\frac{\partial w}{\partial \bar{z}_j}\right|\right\}\right\}$$

q.e.d.

<u>Proof of Theorem 1</u>: We can now complete Step 3.

Let $\Sigma, K, u, U, w_\varepsilon$ be as in Lemma 4. Fix $z_0 \in K$ and put $B_\varepsilon = \{ z \in \mathbb{C}^n \mid |z - z_0| < \varepsilon \}$.

Note that $B_\varepsilon \subset \omega_\varepsilon$ for small ε, for $z \in B_\varepsilon \Rightarrow d^2(z,\Sigma) < \varepsilon^2$ and $\varphi(z) = 1$, so $d^2 - \varepsilon^2\varphi < 0$ at z, i.e. $z \in \omega_\varepsilon$. Now in ω_ε, hence in B_ε,

(*) $\bar\partial w_\varepsilon = \bar\partial U$, by (6) and for $z \in \omega_\varepsilon$

$$\left|\frac{\partial U}{\partial \bar{z}_j}\right| \le C d(z,\Sigma)^{e-1} \quad \text{by Lemma 3. Thus}$$

$$\left|\frac{\partial w_\varepsilon}{\partial \bar{z}_j}\right| \le C d(z,\Sigma)^{e-1} \quad \text{in } \omega_\varepsilon. \text{ Also, as we saw,}$$

$z \in \omega_\varepsilon \Rightarrow d(z,\Sigma) < \varepsilon$. So

$\left|\frac{\partial w_\varepsilon}{\partial \bar{z}_j}\right| \leq C\varepsilon^{e-1}$ in ω_ε. By (7) and (10), then,

and using the fact that $B_\varepsilon \subset \omega_\varepsilon$,

$$|w_\varepsilon(z_0)| \leq C\{\varepsilon^{-n}\cdot O(\varepsilon^{e-1+n-\frac{k}{2}})+\varepsilon\cdot\varepsilon^{e-1}\}.$$

Assume now that $e > \frac{k}{2}+1$. Then

$$|w_\varepsilon(z_0)| = O(\varepsilon^{e-(1+\frac{k}{2})})+O(\varepsilon) = o(1).$$

This $o(1)$ is uniform on K.

Finally $U-w_\varepsilon$ is holomorphic in ω_ε by (*), and

$$U = (U-w_\varepsilon)+w_\varepsilon \text{ in } \omega_\varepsilon.$$

Since $|w_\varepsilon| = o(1)$ uniformly on K, U is approximable uniformly on K by functions holomorphic in ω_ε. But $U = u$ on K. So we are done with the proof of Theorem 1.

<u>Note</u>: "Smooth" as a hypothesis on Σ can thus be taken to mean: Σ is of class C^e with $e > \frac{k}{2}+1$, where $k = \dim \Sigma$.

§ 5. Sketch of Proof of Theorem 2

Let Φ denote the map of $\mathfrak{N}$ into $\mathbb{C}^{2n}$ defined by

$$\Phi(z) = (z,\bar{z}+R(z)).$$

and let Σ be the image of $\mathfrak{N}$ under Φ.

If we assume that R is of class $n+2$ in $\mathfrak{N}$, then Σ is an C^{n+1} submanifold of an open subset of $\mathbb{C}^{2n}$. The dimension of Σ is $2n$, so the smoothness of Σ allows application of Theorem 1 to Σ.

$\Phi(X)$ is a compact subset of Σ. We claim

(i) Σ has no complex tangents

(ii) $\Phi(X)$ is polynomially convex.

Both assertions can be shown to follow from (2). Note that both assertions clearly hold when $R = 0$.

Theorem 1 now gives $P(\Phi(X)) = C(\Phi(X))$ and this is equivalent to

$$[z_1,\ldots,z_n,\bar{z}_1+R_1,\ldots,\bar{z}_n+R_n;X] = C(X).$$

q.e.d.

REFERENCES

[1] F.A. Valentine, A Lipschitz condition preserving extension of a vector function, Amer.J.Math. 67(1945), 83-93.

[2] J. Wermer, Approximation on a disk, Math.Ann. 155(1964), 331-333.

[3] E. Bishop, A minimal boundary for function algebras, Pacific J.Math. 9(1959).

[4] J. Wermer, Polynomially convex disks, Math.Ann. 158(1965), 6-10.

[5] M. Freeman, Some conditions for uniform approximation on a manifold, Function Algebras, Scott, Foresman and Co., Chicago, Ill. (1965), 42-65.

[6] R. Nirenberg and R.O. Wells, Jr., Holomorphic approximation on real submanifolds of a complex manifold, Bull.Amer.Math.Soc. 73(1967), 378-381.

[7] R. Nirenberg + R.O. Wells, Approximation Theorems on differentiable submanifolds of a complex manifold, to be published.

[8] L. Hörmander and J. Wermer, Uniform approximation on compact sets in $\mathbb{C}^n$, to be published.

[9] L. Hörmander, An introduction to complex analysis in several variables, D.Van Nostrand Co., Princeton, N.J. (1966).

[10] L. Hörmander, L^2 estimates and existence theorems for the $\bar{\partial}$ operator, Acta Math. 113(1965), 89-152.

[11] C.B. Morrey, The analytic embedding of abstract real analytic manifolds, Ann.Math. (2) 68 (1958), 159-201.

[12] J.J. Kohn, Harmonic integrals on strongly pseudo-convex manifolds I,II, Ann.of Math. 78(1963), 112-148 and 79(1963), 450-472.

Offsetdruck: Julius Beltz, Weinheim/Bergstr.